Solar & Lunar Eclipses Activity Book

By Kevin Manning

What Causes a Solar or Lunar Eclipse?

What causes a solar or lunar eclipse? When you hear the word eclipse, I want you to think of the word "shadow," because it infers that word. An eclipse is an obstruction of a heavenly body in the sky by its entering into the shadow of another body. So shadow is the key word here that we must keep in mind when discussing eclipses.

Let's begin by going to a STEM hands-on activity I created to demonstrate the importance of the shadow with either a solar or a lunar eclipse. We have three templates we need to preferably print onto heavy card stock of the sun, moon, and earth and cut out using scissors. Once the templates are ready, we can begin doing some simple experiments.

As big as it is, the sun is rounder and more spherical than a bowling ball. We can fit over a million earths inside of the sun! The sun is over a million km in diameter, which is the same as saying it's 865,000 miles in diameter from one end of the sun to the other. Begin cutting out the sun and moon templates around the edges of each. The moon template is a lot smaller than the sun and of course the moon is much smaller than the sun.

Using the sun and moon cutouts you made, hold the sun facing you at arm's length with one hand, and hold the moon facing you a little closer to your eye. Adjust the distance of the moon from your eye until the moon just covers the sun and appears the same size. That's how a total solar eclipse occurs. Now cut out the earth template.

We can fit a million earths inside the sun!

Solar & Lunar Eclipses

What's Coming Up!

<u>Some Definitions for How and When an Eclipse Occurs</u>

1. The obstruction of a heavenly body by its entering into the shadow of another body causes an eclipse.

2. What occurs whenever the moon passes behind the earth such that the earth blocks the sun's rays from striking the moon is a lunar eclipse.

3. What occurs when the moon passes between the Sun and the Earth so that the Sun is fully or partially covered is a solar eclipse.

4. What can occur only when there is a full moon is a lunar eclipse.

5. What phase must the moon be in to experience a total solar eclipse? new moon

Total Solar & Lunar Eclipses
How the Sun & Moon Appear the Same Size in the Sky

Materials: If you can **print out the templates onto heavy card stock**, you will only need scissors, a marble, and a flashlight. If not, then you will also need 1 large paper plate, 2 small paper plates, sun, moon and earth templates, scissors, glue or tape, and a flashlight

Directions:

1. Cut out sun template from printed card stock, or paste or tape to large paper plate
2. Cut out moon template from printed card stock, or paste or tape to small paper plate
3. Cut out earth template from printed card stock, or paste or tape to small paper plate
4. Hold sun upright facing you at arm's length (or tape to a wall), with other hand hold moon upright and move closer or further until the moon just covers the sun (entire plate)
5. With earth facing up flat on table (or taped to a wall), hold a marble (moon) close to the earth while shining a flashlight on the earth and move marble across earth left to right
6. With the moon facing up flat on table (or taped to a wall), shine a flashlight directly on the moon and hold the earth between flashlight and the moon moving earth left to right

The diameter of the sun is about 865,000 miles, and the diameter of the moon is about 2,150 miles. The distance to the sun is about 93 million miles, and the distance to the moon is about 240,000 miles. So even though the sun is about 400 times larger in diameter than the moon, it's also 400 times further away, and the sun and moon "appear" the same size in the sky. That's why we can witness a total solar eclipse at times.

Sun Template

Moon Template

Earth Template

The diameter of the sun is about 400 times wider than the diameter of the moon.

But in the sky, they appear to be the same size.

The sun is also about 400 times further away from us than the moon is, so the average size-to-distance ratios are precisely identical.

So, here's a drawing showing the earth in the middle. This drawing is not to scale at all, but it illustrates a shadow from the earth blocking the sunlight from reaching the moon. When the moon enters the direct shadow known as the umbra, which comes from the English word umbrella, there is a total lunar eclipse happening. The much wider indirect shadow is called the penumbra, and that is where a partial lunar eclipse can be seen. When the moon is on the opposite (night) side of the earth from the sun, the sun sets in the west and the moon rises in the east at the same time, it is a full moon phase. A total lunar eclipse can only happen during a full moon.

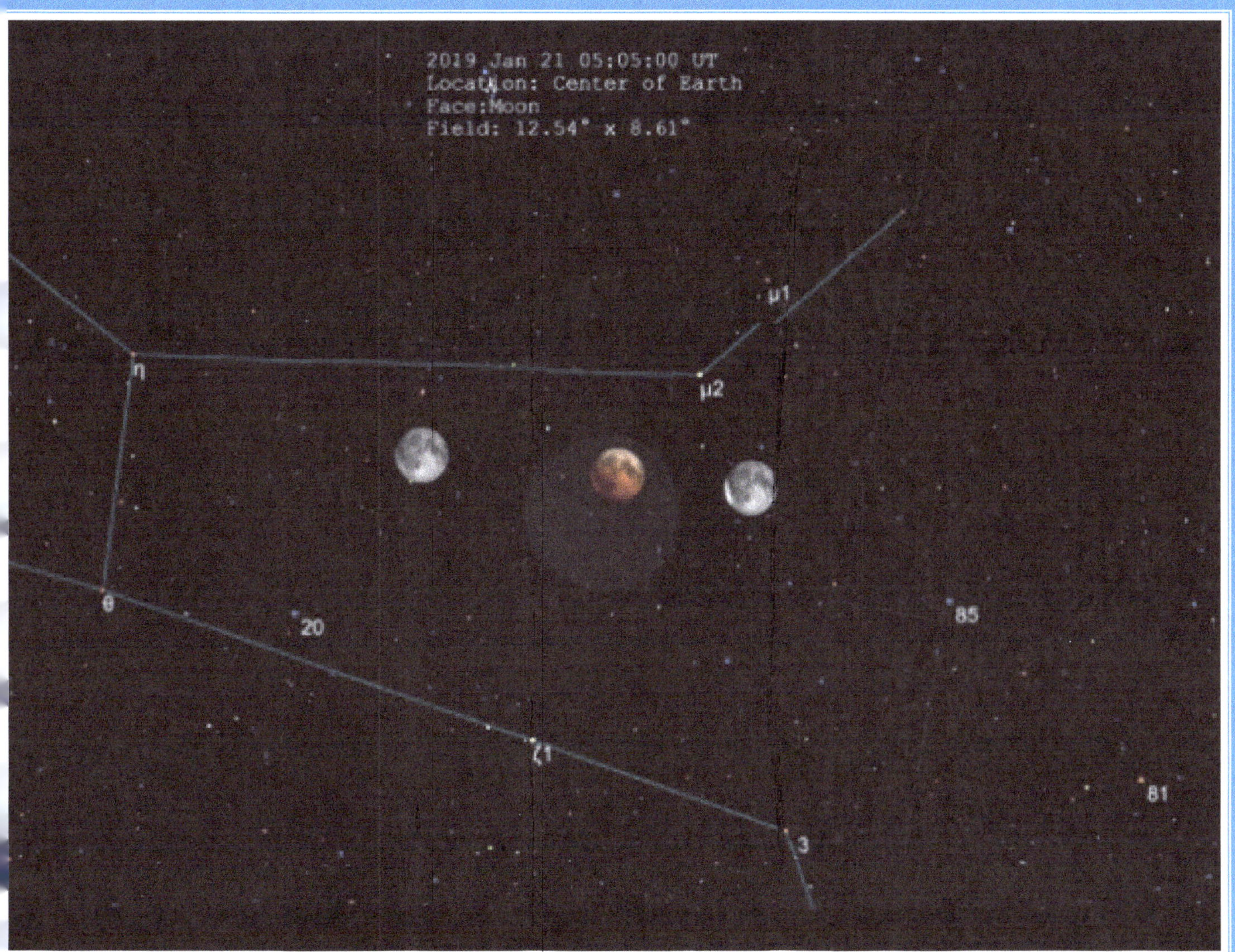

When the moon enters the direct umbral shadow in the sky it appears to change color to red. Some call it a blood moon for that reason. Why does the moon turn red? The sunlight is skimming along the edge of the earth on a line that separates daylight from nighttime, so half of the earth along that circular line is seeing a sunset while the other half sees a sunrise.

Well, you know how the sky appears reddish during both a sunset and a sunrise? All that red sky is being projected onto the moon, and that's why the moon appears red during a total lunar eclipse.

So, with a total lunar eclipse, we begin with the earth blocking the sunlight from reaching the moon just a bit and as it progresses, we see less and less sunlight still reaching the moon. Along the shadow line on the moon, we notice that it is curved. That's because the earth is round and somewhat spherical like a ball, and even though the earth is rotating or spinning the curved shadow stays constant.

Here's a montage of several photos of the moon before, during, and after a total lunar eclipse as it turns red when it enters the umbra shadow from the earth. From the beginning of a partial eclipse through totality and to the end of the partial on the other side it can take an average of about 3 hours or more. Totality can last 100 minutes.

The recent total lunar eclipse seen on May 15-16, 2022, had a duration of totality close to 90 minutes.

Super Blood Wolf Moon January 20-21, 2019

Here's a closeup view through my telescope of a total lunar eclipse on January 20-21, 2019, called a Super Blood Wolf Moon. Super because it was closer to the earth in its orbit and appeared larger than average, Blood because of the red color, and Wolf because Native American Indians long ago would hear the wolves cry at night a lot during the cold winter of January.

Set the moon cutout flat on a table facing up towards you and shine a bright flashlight (representing the sun) onto the moon while slowly moving the earth cutout between the flashlight and moon so its shadow starts at one end and goes across the moon off on the other side. This activity simulates a total lunar eclipse.

Then set the earth cutout face up on the table shining the flashlight on it while slowly moving a marble a couple of inches above the earth so the marble's shadow starts at one end of the earth and goes across to the other end. This activity simulates a total solar eclipse.

Again, not to scale, let's look at this drawing of where the shadow forms from the moon in the middle between the sun and the earth. The direct umbra shadow from the moon blocking the sunlight reaching the earth can vary between 62 and 165 miles wide with an average less than 70 miles. The secondary penumbra shadow where a partial eclipse is seen, however, can span a diameter of nearly 4,000 miles across.

Here we see the real umbra shadow upon the earth during a total solar eclipse as photographed by a French astronaut on the Mir Space Station in 1999.

Conditions Favorable for an Eclipse

Why don't eclipses occur every month? The moon's orbit around the earth is tilted a little over 5 degrees compared to the earth's orbit around the sun. Because of this tilt, the line of nodes as we call them don't line up so that the shadows make contact. At two points during our annual journey around the sun the line of nodes points toward the sun and it's possible for an eclipse to occur. These are known as eclipse seasons. On average, a total lunar eclipse can happen for any given location once every 2 1/2 years. For a total solar eclipse, it can happen on average every 1 1/2 years or 18 months. Of course, that means an eclipse seen somewhere in the world, not necessarily where you live.

This is a real photo of a total solar eclipse. The black ball you see in the middle is the moon. Behind the moon the sun, much further away, is playing peek-a-boo. The moon appears large enough to cover the entire sun behind it. The beautiful butterfly shaped glow is the sun's atmosphere known as the solar corona, which can be seen during totality. The shape of the corona is determined by the magnetic field lines coming up from the core of the sun. Also, during totality, it looks like a sunset all the way around along the horizon with an eerie silvery hue higher up that looks otherworldly. Even though it's the middle of the day it gets dark out, dark enough to see planets and bright stars in the sky. The temperature can drop 10, 20, even 30 degrees reaching the condensation point in the atmosphere causing the moisture to condense into water drops, so even though there's not a cloud in the sky it can begin raining. Nearby animals you heard before and after the event get quiet because they think it's nighttime. Nocturnal insects such as crickets are making their clicks instead.

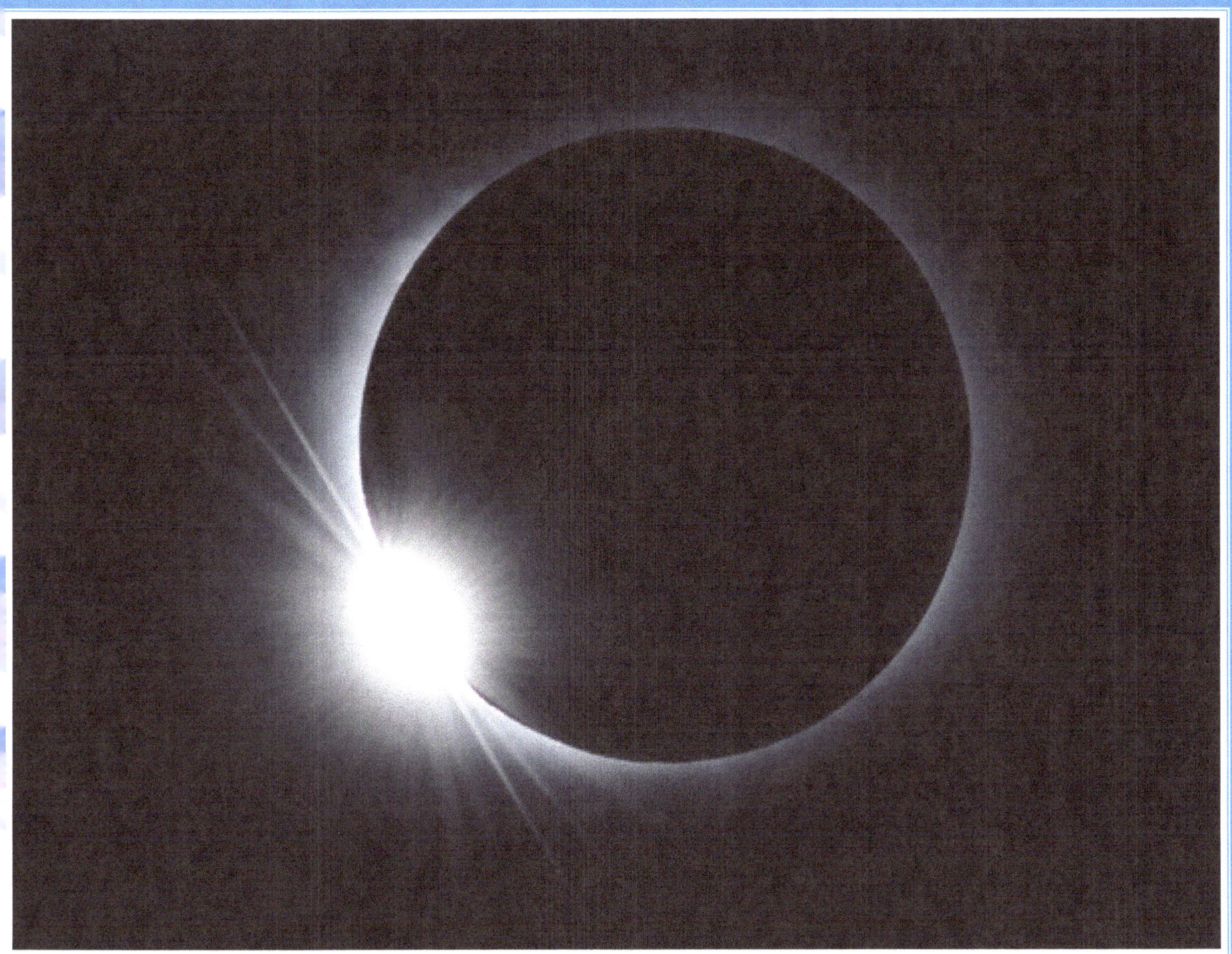

Following totality when the moon just begins to uncover a small sliver of sunlight, a bright bulge develops with a dim glow around the moon forming what is commonly referred to as the "diamond ring effect." Can you see the diamond ring?

This view shows the progression of the eclipse from beginning to end, with the middle view in totality and the solar corona shows. The moon is always moving as it orbits around the earth.

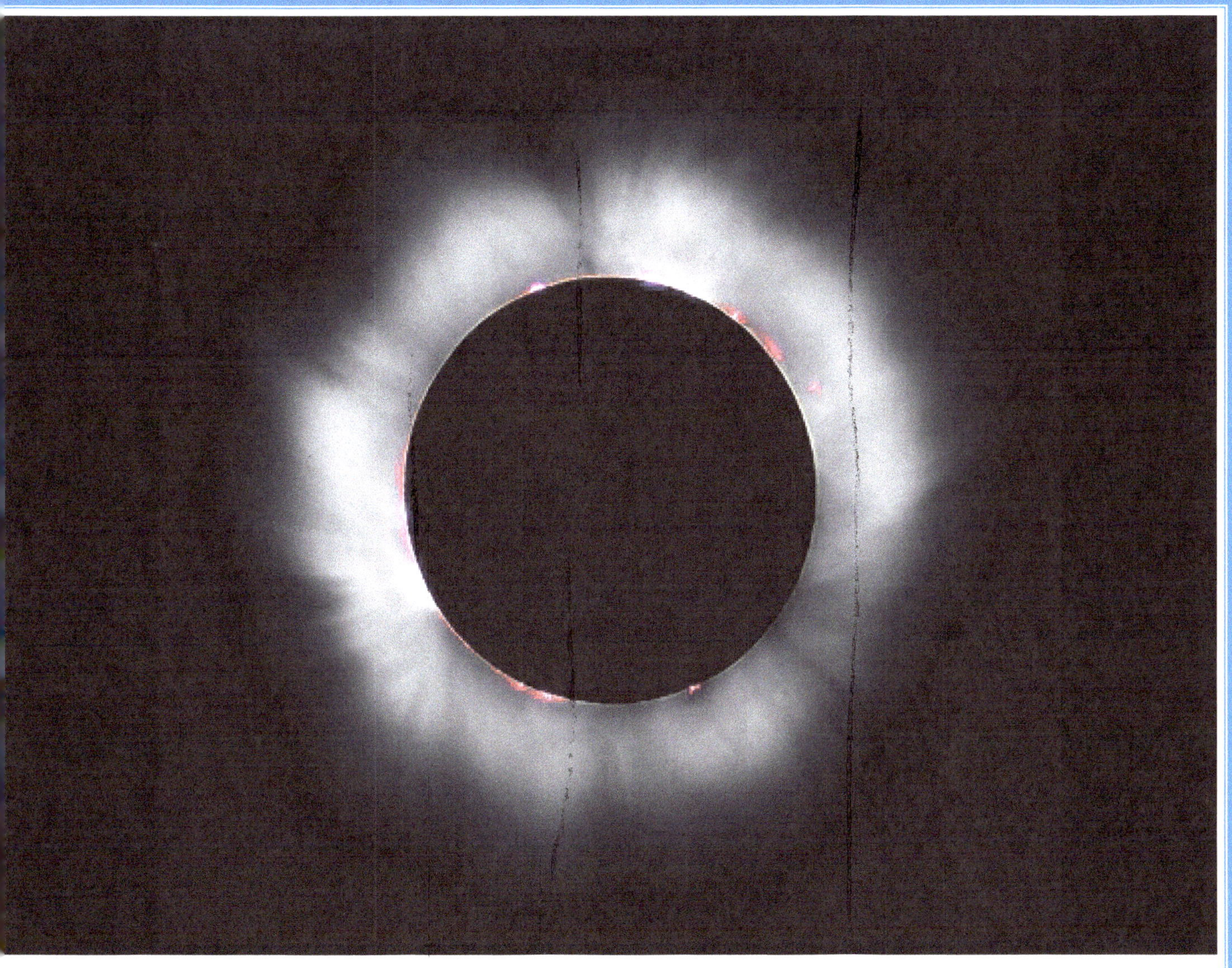

Here we see what looks like little red flames arising from the solar surface of the sun, ejections of the 4th phase of matter called plasma that the sun is made of along with superheated gas. Known as solar prominences, these areas are many times larger than the earth and are also associated with the magnetic field of the sun.

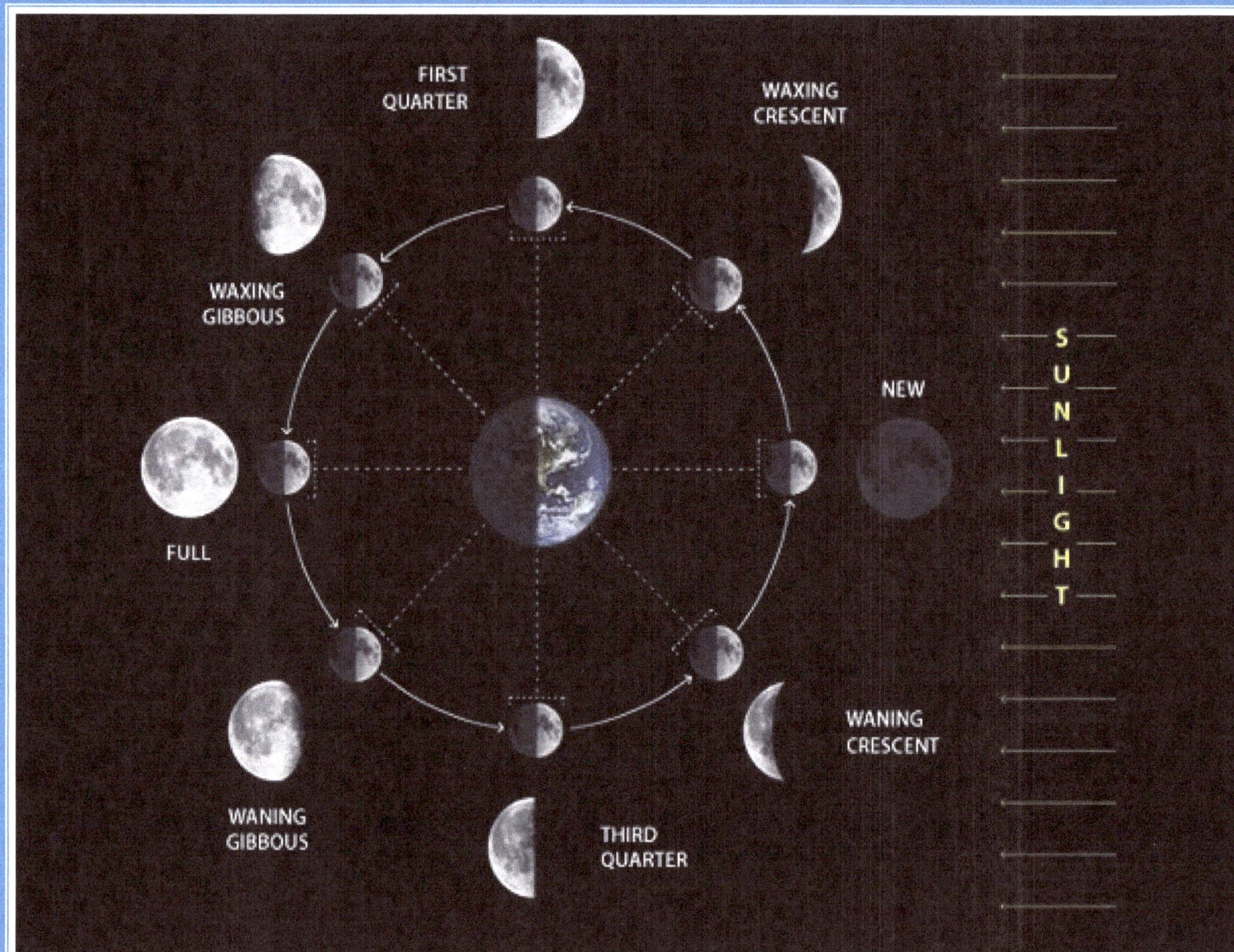

The moon orbits the earth in one moonth! I'm joking, but that's where we get the word month from. This one synodic month is about 29 1/2 days. As the moon orbits us it goes through these phases from our perspective here on the earth. If we begin with the moon facing toward the direction of the sun it is a new moon phase. This is when a total solar eclipse can happen. As it goes around the earth it appears as a crescent phase and is waxing, meaning that there's more sunlight seen on the moon and to looks larger. One-fourth the way around, the moon appears half lit by the sun and it's first quarter phase. I'd like to point out that if it wasn't for the sun and its bright light, we wouldn't be able to see the moon or planets at all. They don't give off their own light like the sun and other stars do. As the moon progresses in its orbit, we come to a waxing gibbous where the moon is between half lit up and full. Then at full phase the moon is on the other side of the sun and this is when a total lunar eclipse can occur. From here on the amount of sunlight reflecting off the moon grows smaller each night, so we see a waning gibbous, third or last quarter, and a waning crescent before we get back to new moon and the cycle starts all over.

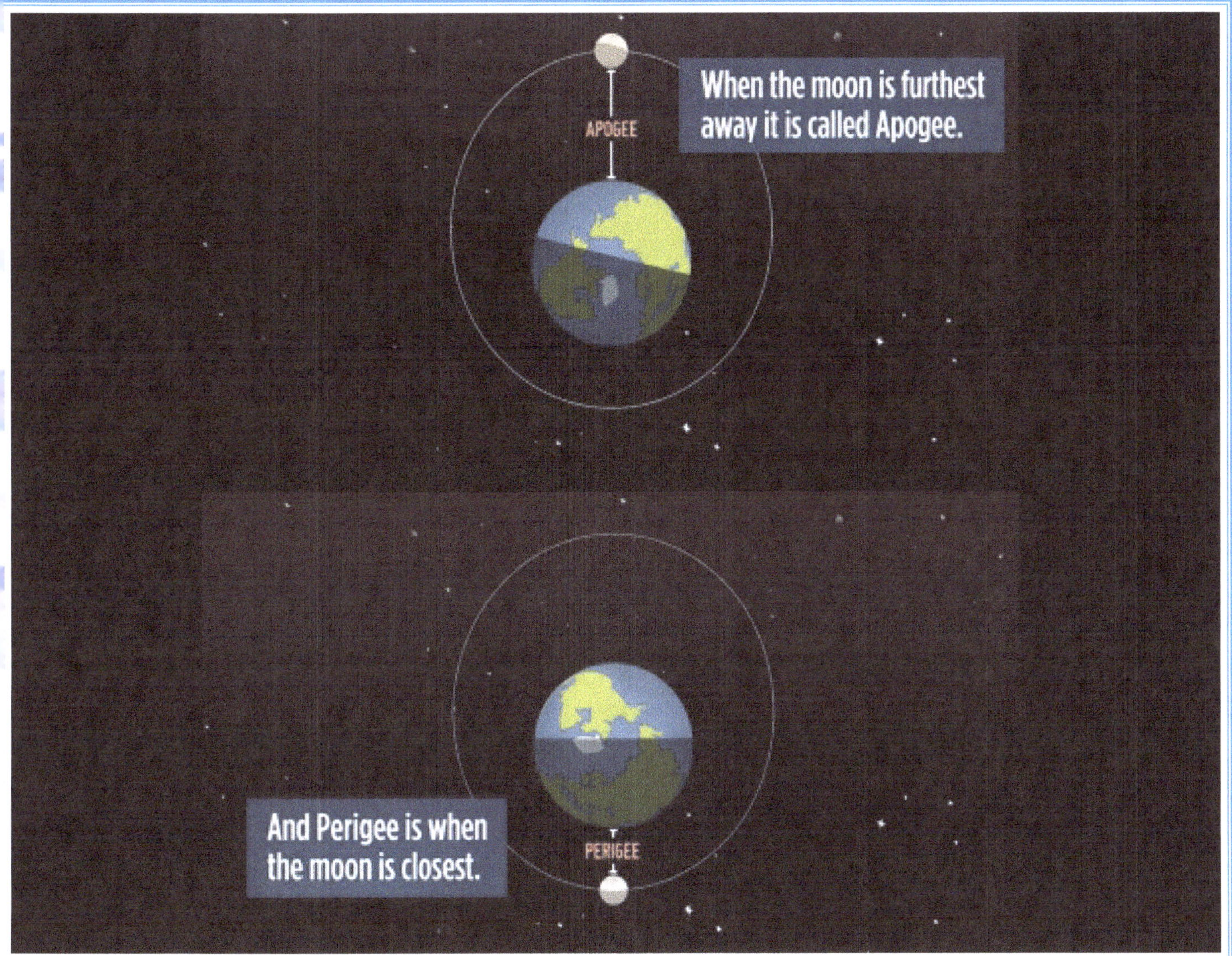

Now as the moon orbits the earth it doesn't trace out a circular path but more of an cllipsc. That oval shape makes the moon travel further from the earth at times called apogee, and at other times get closer to the earth called perigee.

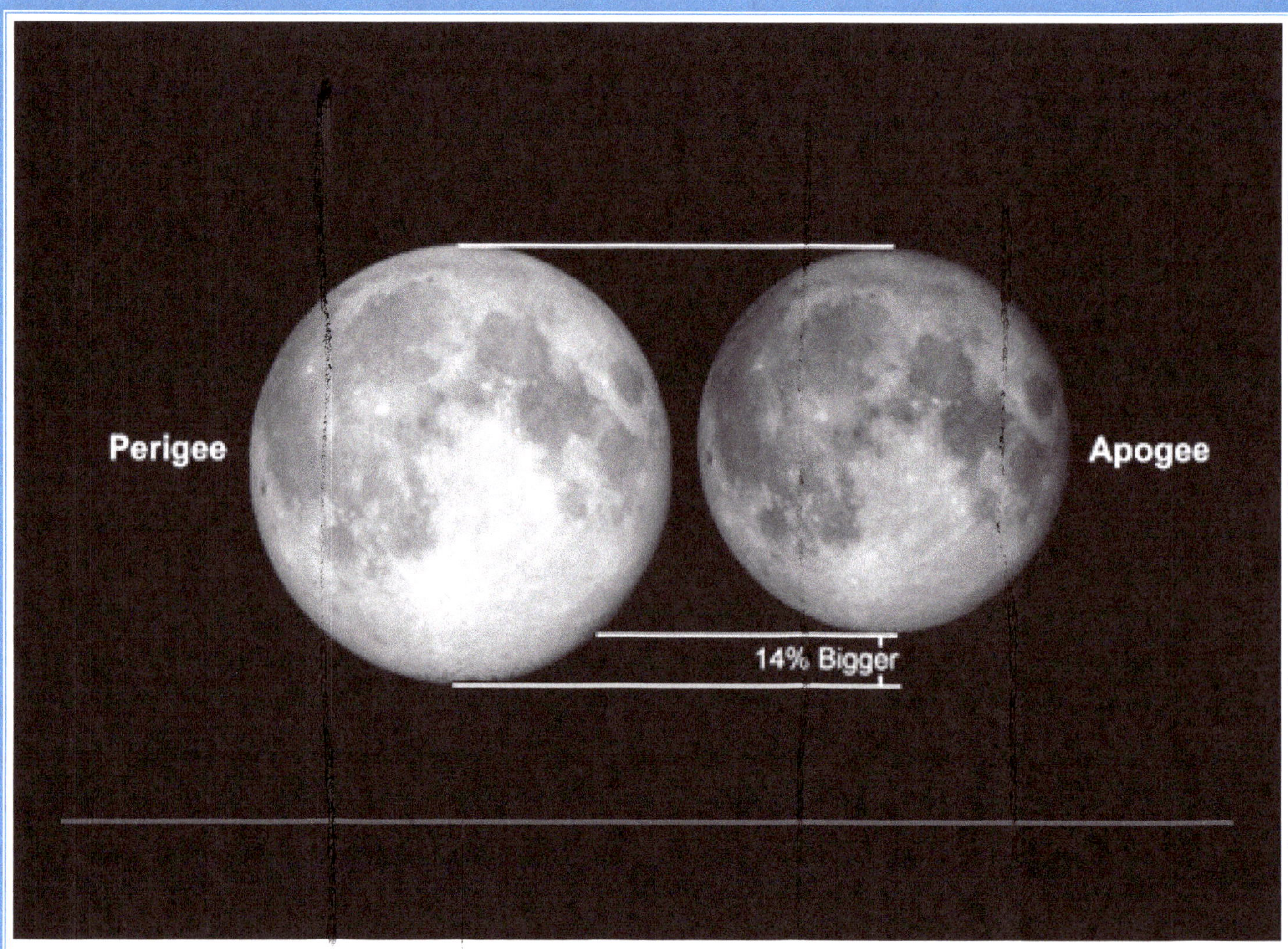

When the moon is at perigee it appears about 14% larger and about 30% brighter in the sky. We call this a Super moon. Many people mistakenly think the moon is many times larger when it's near the horizon, but that is just an optical illusion because it's near terrestrial objects like buildings and trees compared to nothing nearby when the moon is high in the sky.

This photo shows the comparison of the moon at apogee vs at perigee. So, what happens when the moon is at apogee during a solar eclipse? Well, it's not quite big enough to cover the entire sun behind it and when centered leaves an annulus ring of fire or the rim of the sun surrounding the moon. This kind of event only happens once every 360 years at any particular geographic location, so if you can catch one in your lifetime you're doing well.

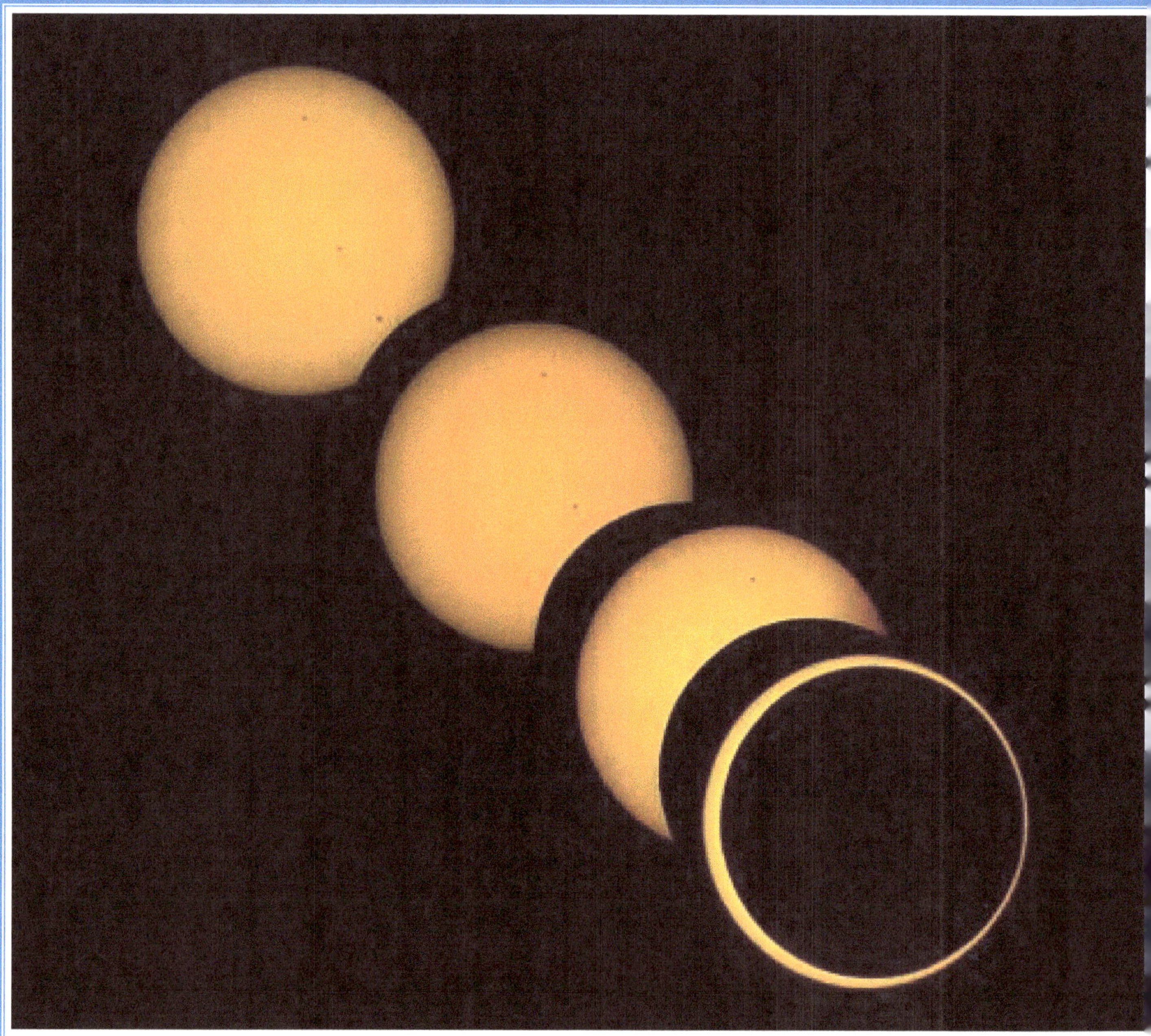

Here's a view of the progression during different time frames of an annular solar eclipse.

Back to our sun and moon cutouts, we saw that the distance the moon is from the earth, about 240,000 miles on average, makes it appear around the same size as the sun's disk in the sky. If the moon is a bit closer during an eclipse, it will appear slightly larger and keep the sun covered for a longer period. But what happens when the moon is further away and appears a little smaller than the sun? Then we have a rarer event known as an annular solar eclipse.

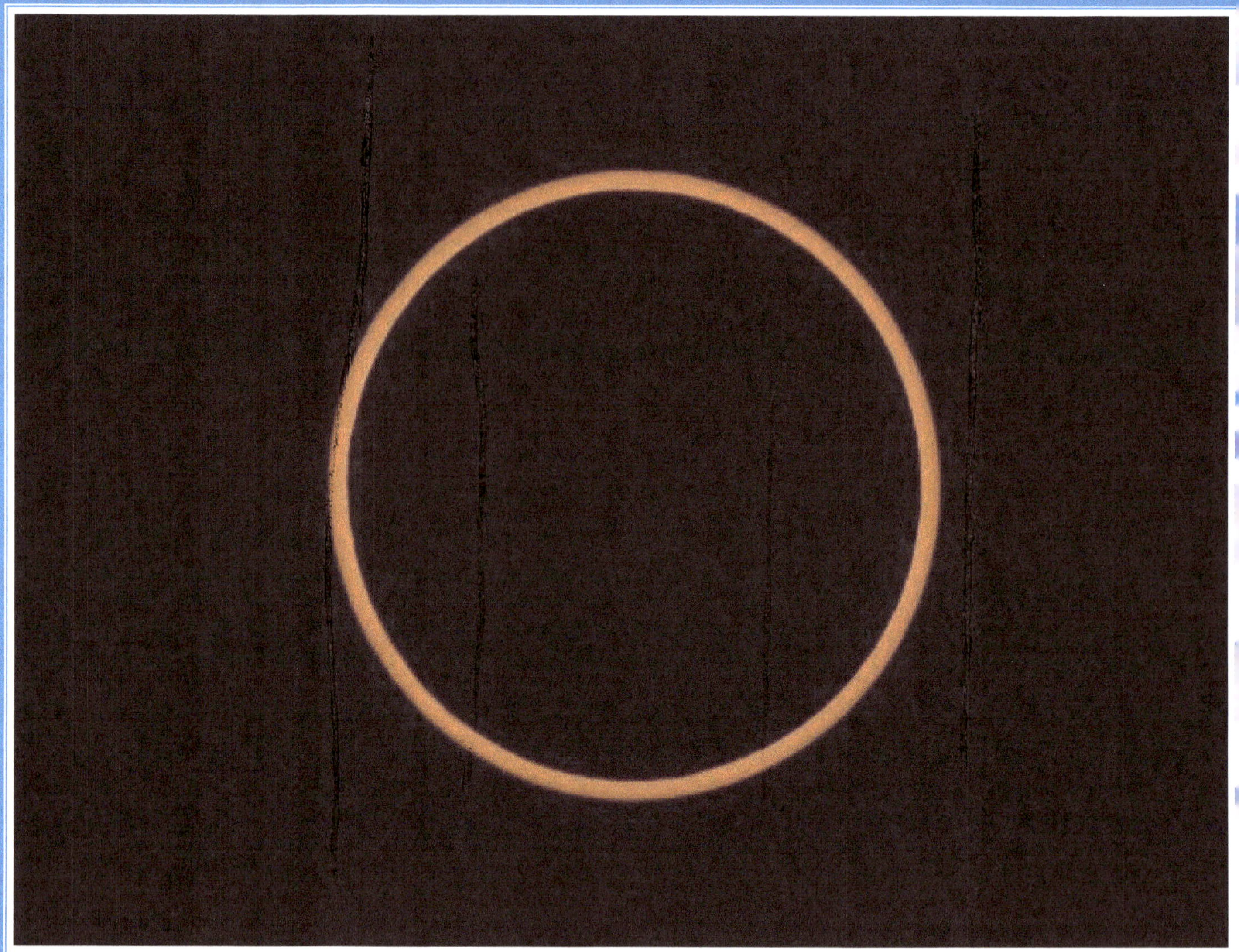

I was fortunate in seeing an annular solar eclipse and took this photo when the moon was dead center in the middle of the sun using a solar filter, and that's why it looks orange.

Types of Solar Eclipses

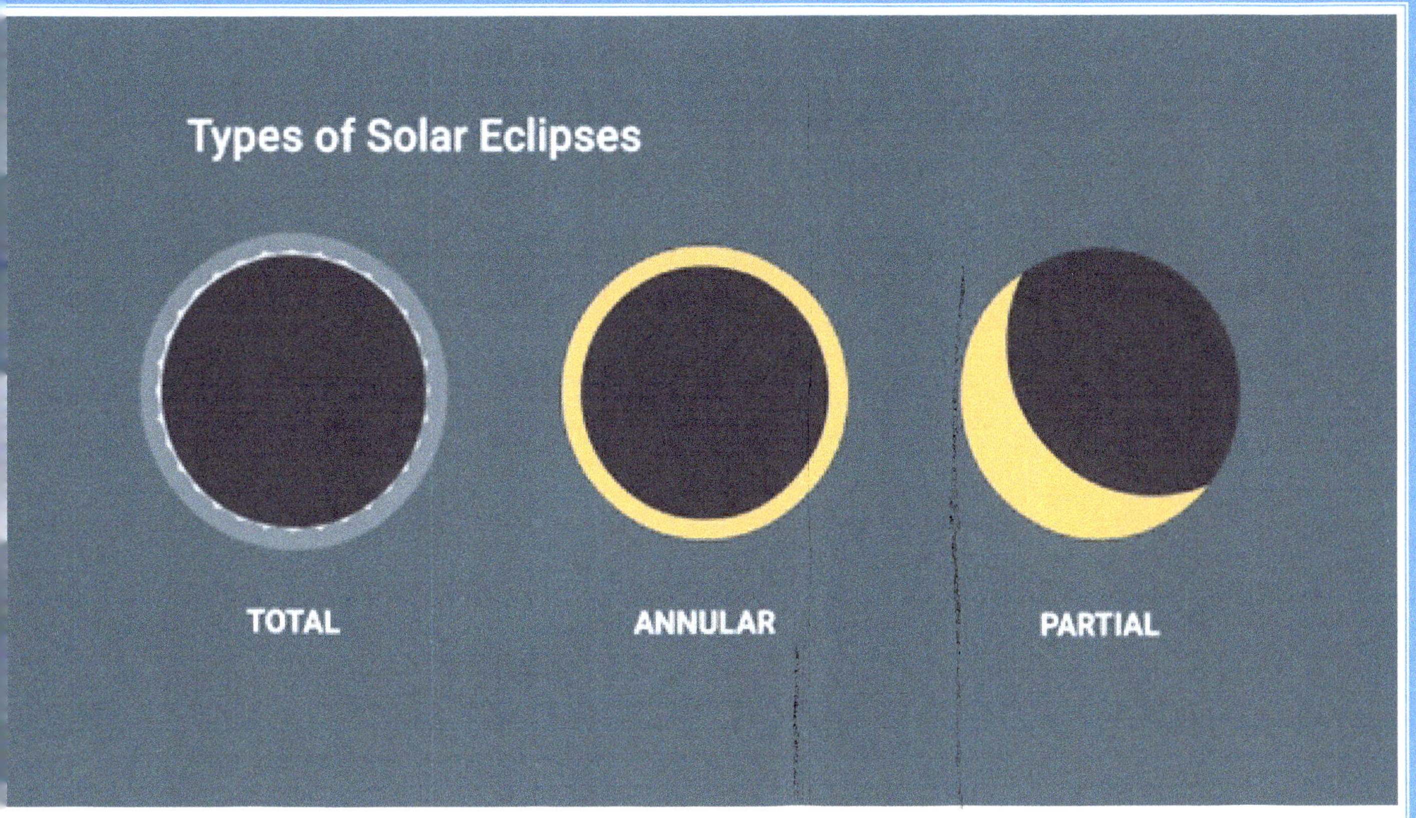

There are 3 types of solar eclipses: total, annular, and partial.

As you may recall earlier, the penumbral shadow where a partial eclipse occurs is a much larger geographic area on the earth and therefore more common and seen by many, many more people in the world.

Here's a fisherman near the seashore and seagulls in flight where a partial eclipse is seen near the horizon.

A jet aircraft is seen in this photo of a slightly partial eclipse.

The tiny spaces between leaves on a tree can form small pinholes and like a pinhole camera will show many focused images of the event on the ground or the petal of a flower.

Projection of an eclipse image through a telescope can be the safest method used as long as you don't look through the telescope without a solar filter for protection.

A white cardboard screen can be attached where the focused image forms to see a reflection of the eclipse.

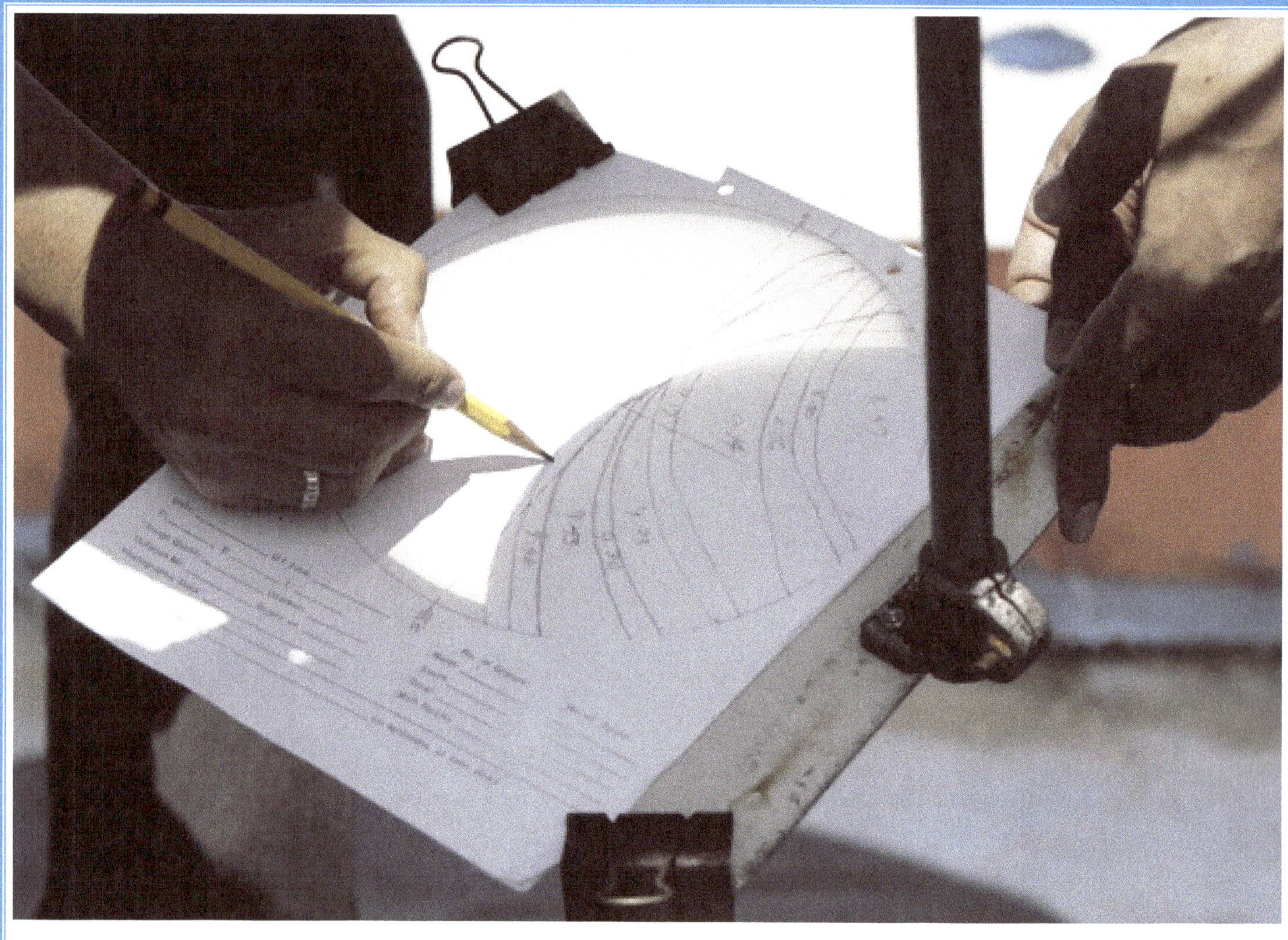

Drawing along the shadow lines with precise timing can be useful scientifically for recording different stages of the eclipse.

Special filters can be used to look directly at the eclipse.

Solar filters come in all shapes and sizes.

Solar filters are available for telescopes that only transmit 1/100,000th of the suns bright light and blocks the harmful rays for a closeup view of the eclipse and even sunspot groups on the sun's surface.

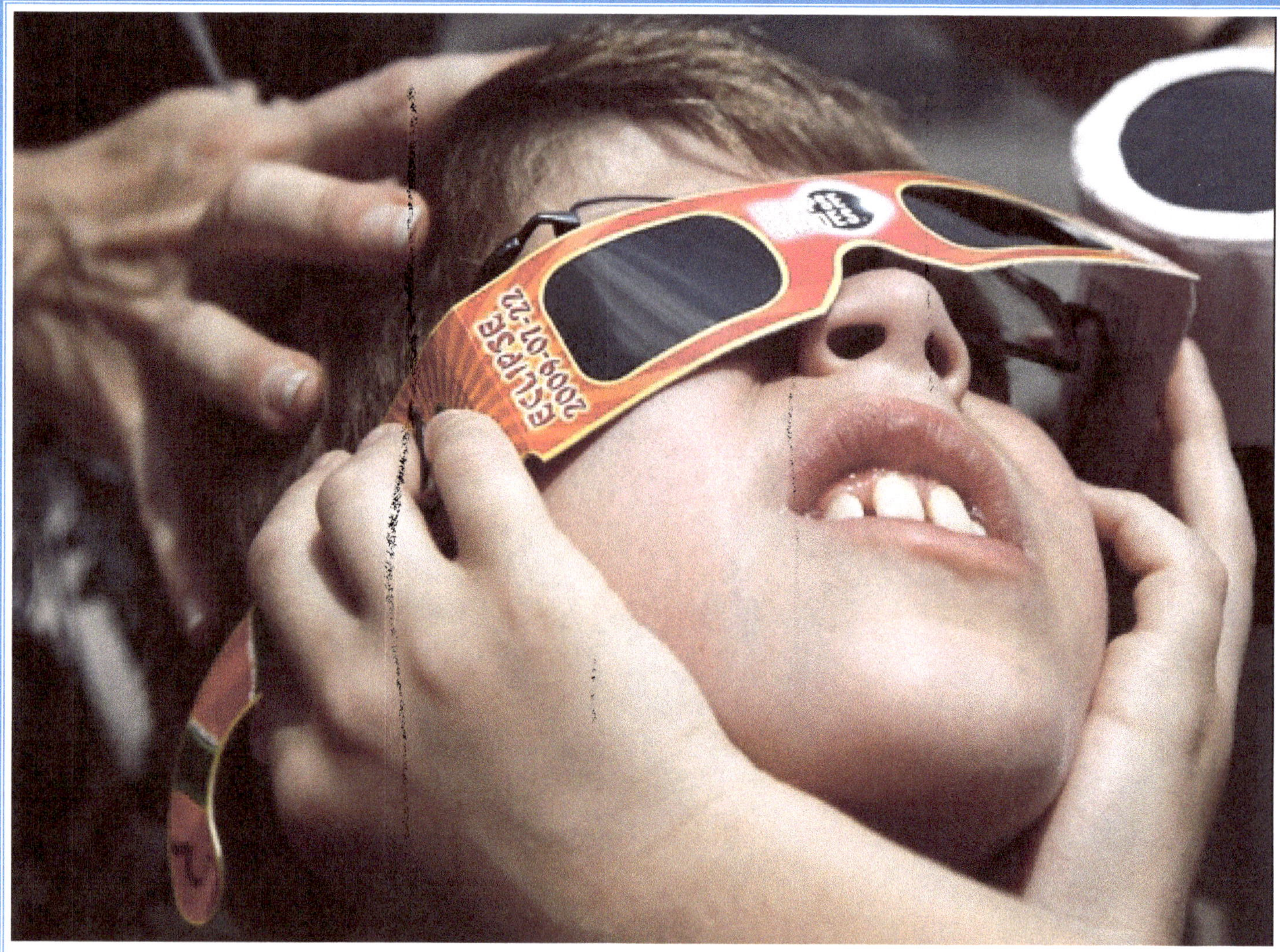

For years now, many people get eclipse glasses to wear just like regular glasses but have a small solar filter for each eye.

These special glasses are available everywhere…..

…..around the world!

I don't recommend welder's filters as they don't block some of the harmful rays of the sun from getting through, so don't use them.

On August 21, 2017, there was the last total solar eclipse seen across America. I got to see the event while showing it to hundreds of people after doing a NASA livestream earlier that day.

I have some very exciting news! We have a couple of major events coming up in the next two years. On October 14, 2023, there will be that rare annular solar eclipse type crossing much of the western United States. The following year on April 8, 2024, there will be a total solar eclipse across much of the eastern United States. The shaded areas show the paths of totality where the moon will be centered in front of the sun with a ring of fire on one and a total obstruction on the other with the solar corona showing. Anyone who lives within the path doesn't need to go anywhere as the eclipse will come right over their location. The two paths cross each other in Texas near the city of San Antonio, so they can witness both events right where they live. The rest of us will need to travel to where the paths are and trust me as one who has seen a few of these events that it may be well worth the effort. Just make sure the weather forecast for where you are traveling to is favorable with no clouds and clear skies for the most part. Otherwise, your long trip could turn out to be in vain. For those who get to see these types of events unhindered by clouds it can be a lifechanging experience never forgotten.

If you are not able or willing to travel to the paths, keep in mind that all the rest of the United States and beyond will experience a partial solar eclipse for both events.

Use Safe Eclipse Glasses!

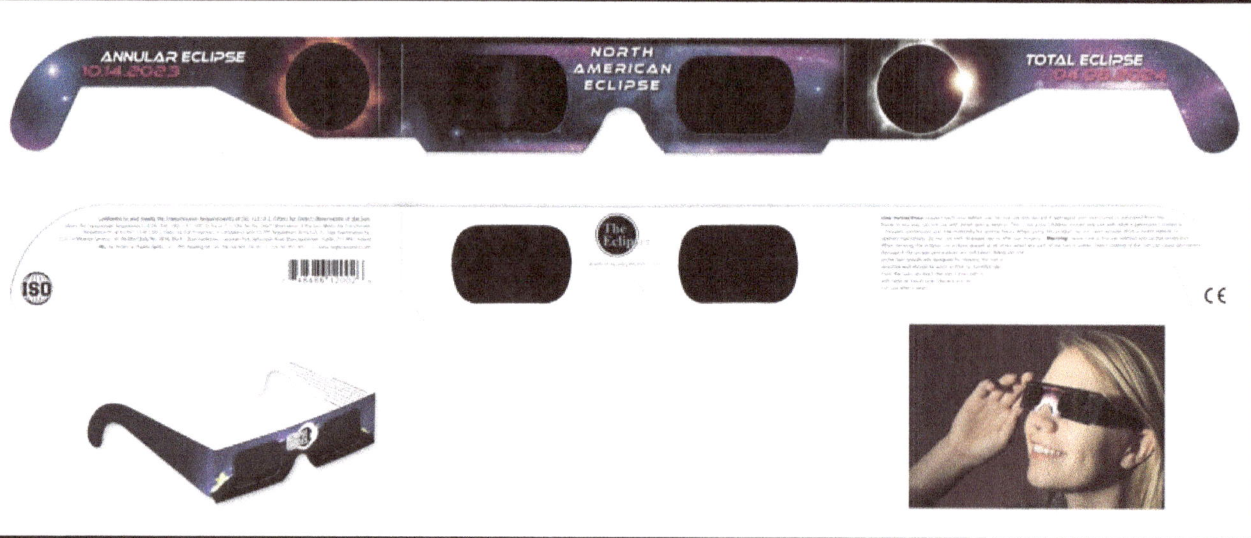

Only Use Glasses that are Tested, Approved and Certified

https://www.eclipseglasses.com?sca_ref=1941651.fGjEPtp7qC

Eclipse glasses are available and are very inexpensive especially well in advance of the events. As the time draws near supply may not meet demand and the prices will increase accordingly. The main thing is you want to be sure that the glasses you are acquiring are approved and certified to be safe, so a reputable dealer in the United States may be the way to go.

Many schools, museums, churches and other civic organizations buy many hundreds or even thousands of the glasses at a much-reduced price and then sell them at an increase for a profit as a fundraiser for a good cause.

I know the glasses I get myself are well-made and totally safe, and this is the website you can find them if you wish - https://www.eclipseglasses.com?sca_ref=1941651.fGjEPtp7qC.

Spherical or positional astronomy is the branch of astronomy that is used to determine the location of objects on the celestial sphere, as seen at a particular date, time, and location on the Earth. It relies on the mathematical methods of spherical geometry and the measurements of astrometry.

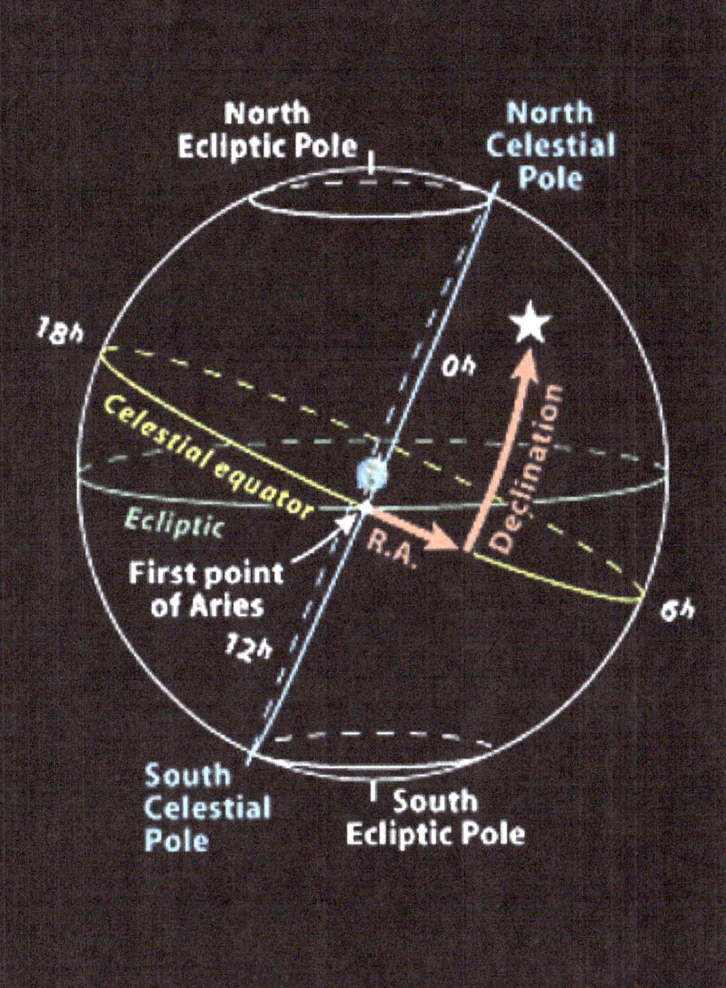

In a 2-dimensional thinking, we can imagine the entire night sky forming this gigantic ball known as the celestial (sky) sphere. It's like we're standing on little blue marble earth looking up from horizon to horizon. In this view, the entire celestial sphere serves as a backdrop where all the stars appear to be fixed.

Long ago we mapped out the whole world using fictitious lines that cris-cross each other. If we measure between lines going around the world east and west from the equator to both poles, we are moving north and south. These lines are called latitude. Measuring between lines appearing to go north and south like peals of an orange meeting at the poles moves us east and west. These are called lines of longitude. Latitude and longitude are what we use in our GPS systems anywhere on the planet. What if we took the lines of latitude and longitude and threw them up into the sky? That's exactly what was done, only we changed the names to protect the innocent. So now astronomers have a roadmap to the stars where lines intersect pinpointing the accurate position of stars and any celestial object in the sky. The sun rises in the east every morning and sets in the west every evening because the earth is spinning or rotating that direction. One day, or about 24 hours, is how long it takes to rotate once in a complete circle. Notice that we are measuring time when we move east-west. A whole day is divided into 24 hours. Each hour is divided into 60 minutes, and each minute is divided into 60 seconds. Why 60? It's because a long time ago we adopted ancient Babylonian mathematics, which is a base-60 arithmetic. Longitude in the sky is called Right Ascension, or abbreviated "RA" for short. Why is it called "Right" Ascension? Any guesses? Remember that when we face north that east is to our right, and the eastern sky is where the sun and everything else in the sky appears to rise, or ascend. Latitude in the sky is called Declination (abbreviated Dec for short), which measures north or south of the celestial equator, or zero degrees.

Since north and south does NOT measure time, we use degrees of arc, the angular parts of a circle like slices of a pie or numbers on a clock. There are 360 degrees in a circle. If we begin with the celestial equator at 0 degrees, then the north celestial pole is plus 90 degrees, or +90. The plus sign means north. The south celestial pole is at minus 90 degrees, or -90. The minus sign means south. In order to pinpoint the position of a celestial object accurately, we need to divide each degree into smaller increments. Since we adopted ancient Babylonian math using Right Ascension, we kept the same base-60 math here as well even though it's not measuring time at all. Each degree of Declination is divided into 60 minutes of arc, or simply 60 arcminutes, abbreviated 60' like a foot sign in English.   Each arcminute can be subdivided into 60 arcseconds, abbreviated 60" like an inch sign in English. To give you some perspective on this, the apparent diameter of the Moon in the sky is 1/2 degree, or 30', or 1,800". It's easy to see that one arcsecond is a very small piece of sky.

Appulse is an astronomical term that refers to the very near approach of one celestial object to another, as seen from a third body. **_Usually_** it refers to the close approach of two planets together in the sky, or of the Moon to a star or planet as the Moon follows its monthly orbit around Earth, as seen by an observer located on Earth. An "appulse" can also be referred to as a conjunction. Where the celestial bodies come so close together that one actually passes over the other, the event is known as an occultation.

The Moon and Venus

Conjunction: The point at which two celestial bodies simultaneously cross a given line

…By definition, this line is either Right Ascension or Ecliptic Longitude.
…When the Moon is new, it is considered to be in conjunction with the Sun.

Venus and Jupiter

Comet Holmes and the Andromeda Galaxy

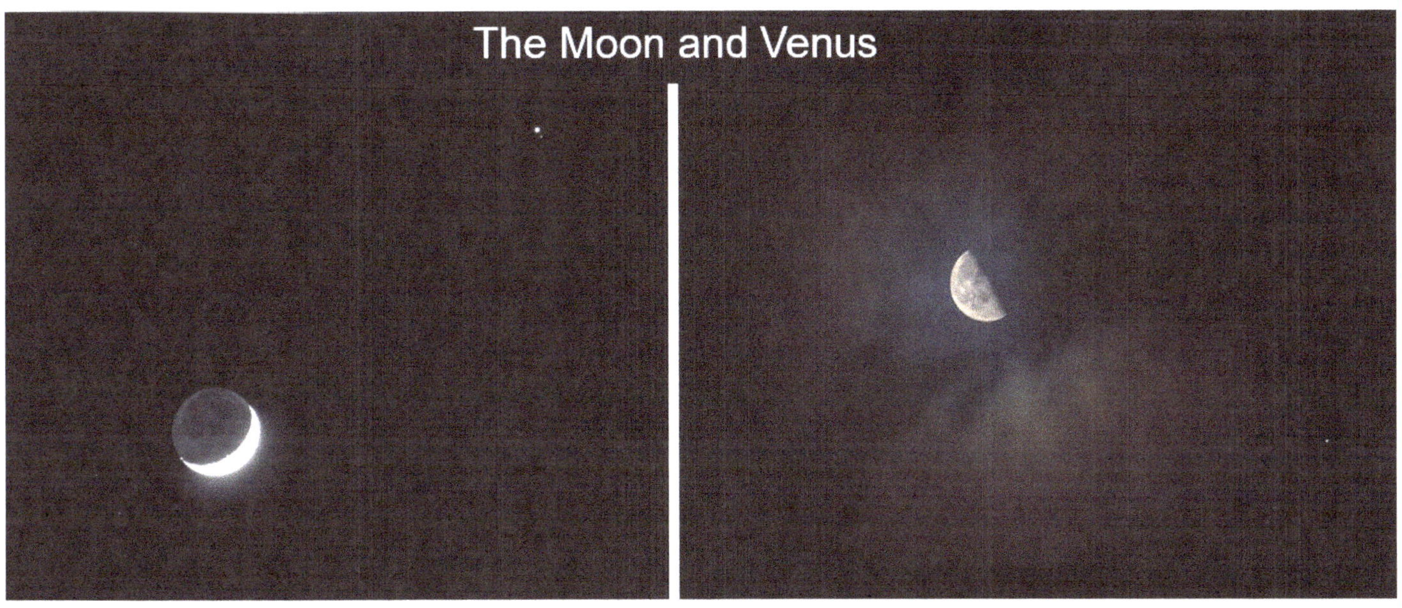

The Moon and Venus

Moon, Venus and Jupiter

Moon, Mercury and Pleiades

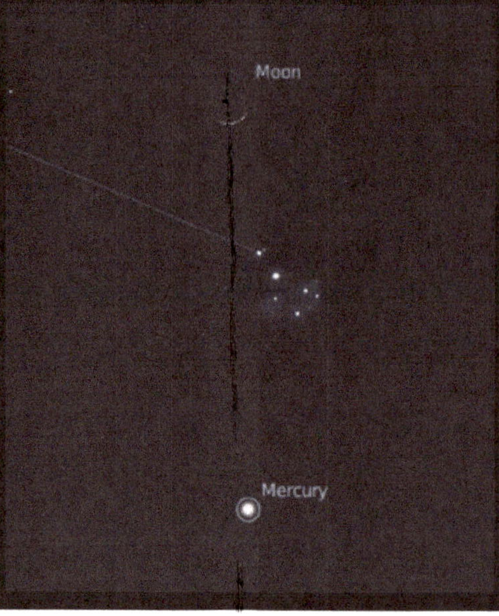

Moon

Mercury

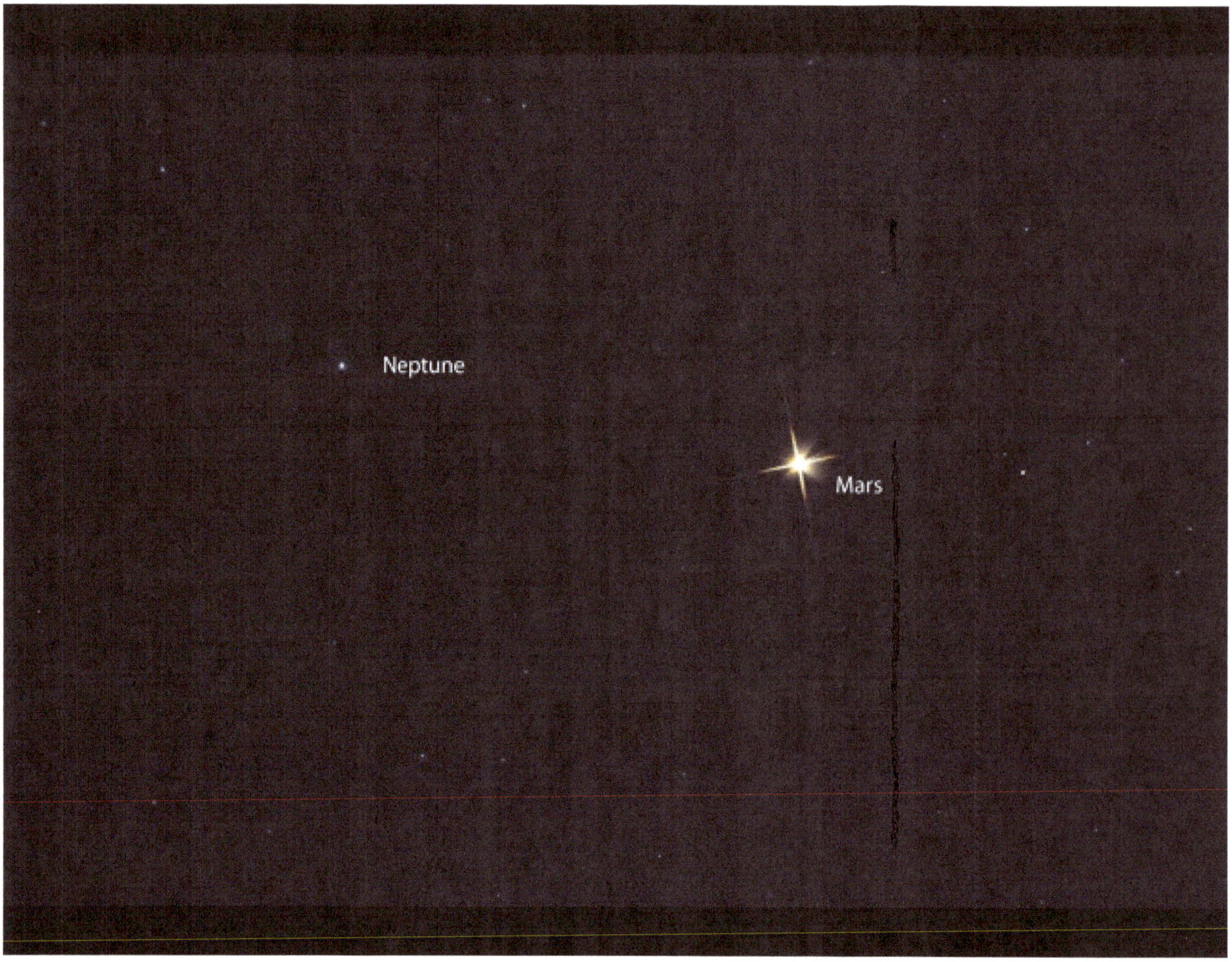

Neptune

Mars

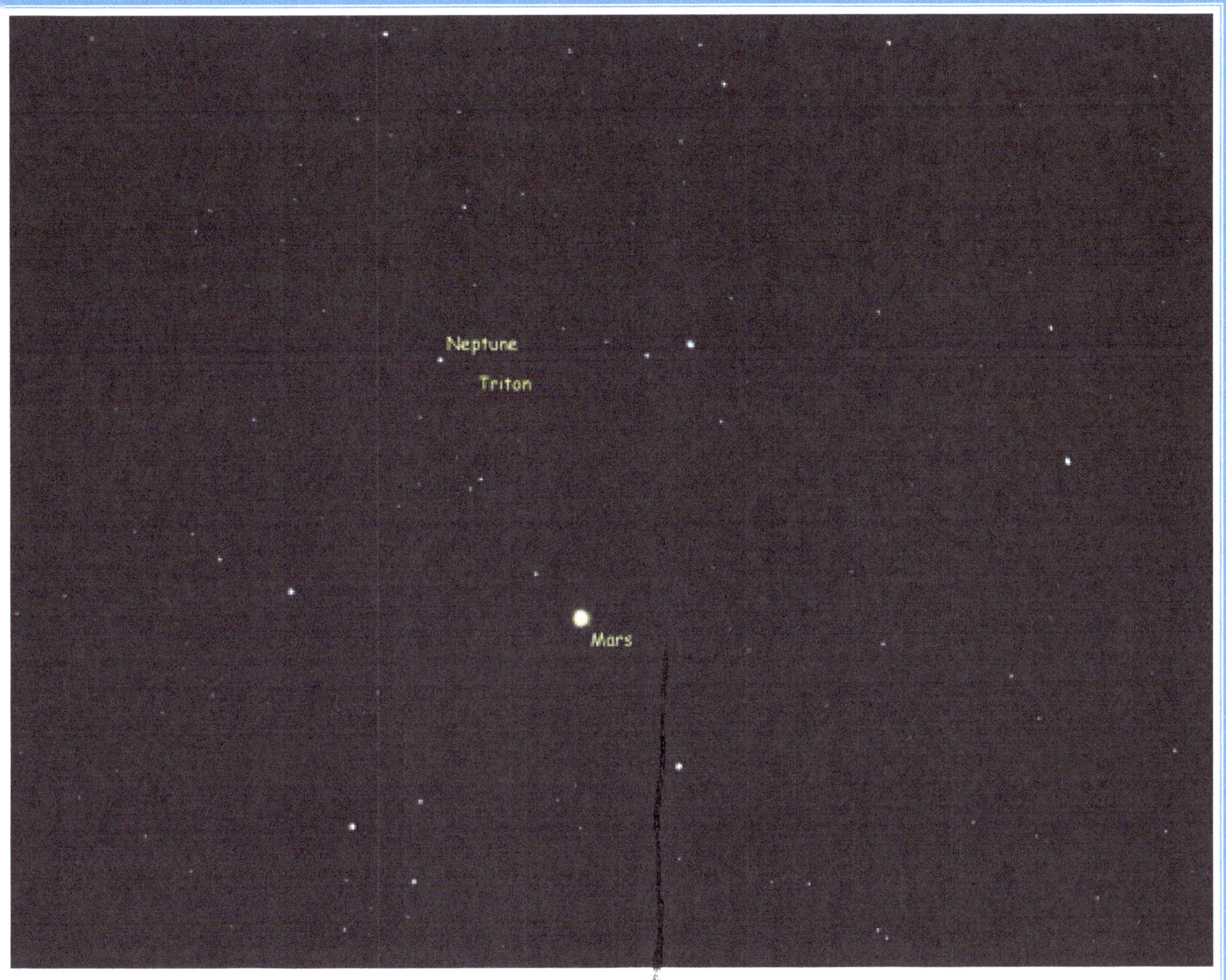

Neptune is just a little smaller than Uranus at 30,775 miles in diameter, the smallest of the gas giant planets. Officially the most distant major planet in our solar system, Neptune is 2.793 billion miles from the sun and takes nearly 164 years to orbit around it. Neptune has a total of 14 moons, and is famous for its large dark spot, a storm with winds approaching 1,500 mph.

Triton was discovered on Oct. 10, 1846 by British astronomer William Lassell, just 17 days after Neptune itself was discovered. Triton is the largest of Neptune's 14 moons. It is unusual because it is the only large moon in our solar system that orbits in the opposite direction of its planet's rotation—a retrograde orbit. Triton has a diameter of 1,680 miles (2,700 kilometers). Spacecraft images show the moon has a sparsely cratered surface with smooth volcanic plains, mounds and round pits formed by icy lava flows. Triton is one of the coldest objects in our solar system. Triton's thin atmosphere is composed mainly of nitrogen with small amounts of methane.

An ___occultation___ is an event that occurs when one object is hidden by another object that passes between it and the observer, taken from the Latin occultare, which means "to conceal." The word is used in astronomy and can also be used in a general (non-astronomical) sense to describe when an object in the foreground occults (covers up) objects in the background.

The Moon and Mars

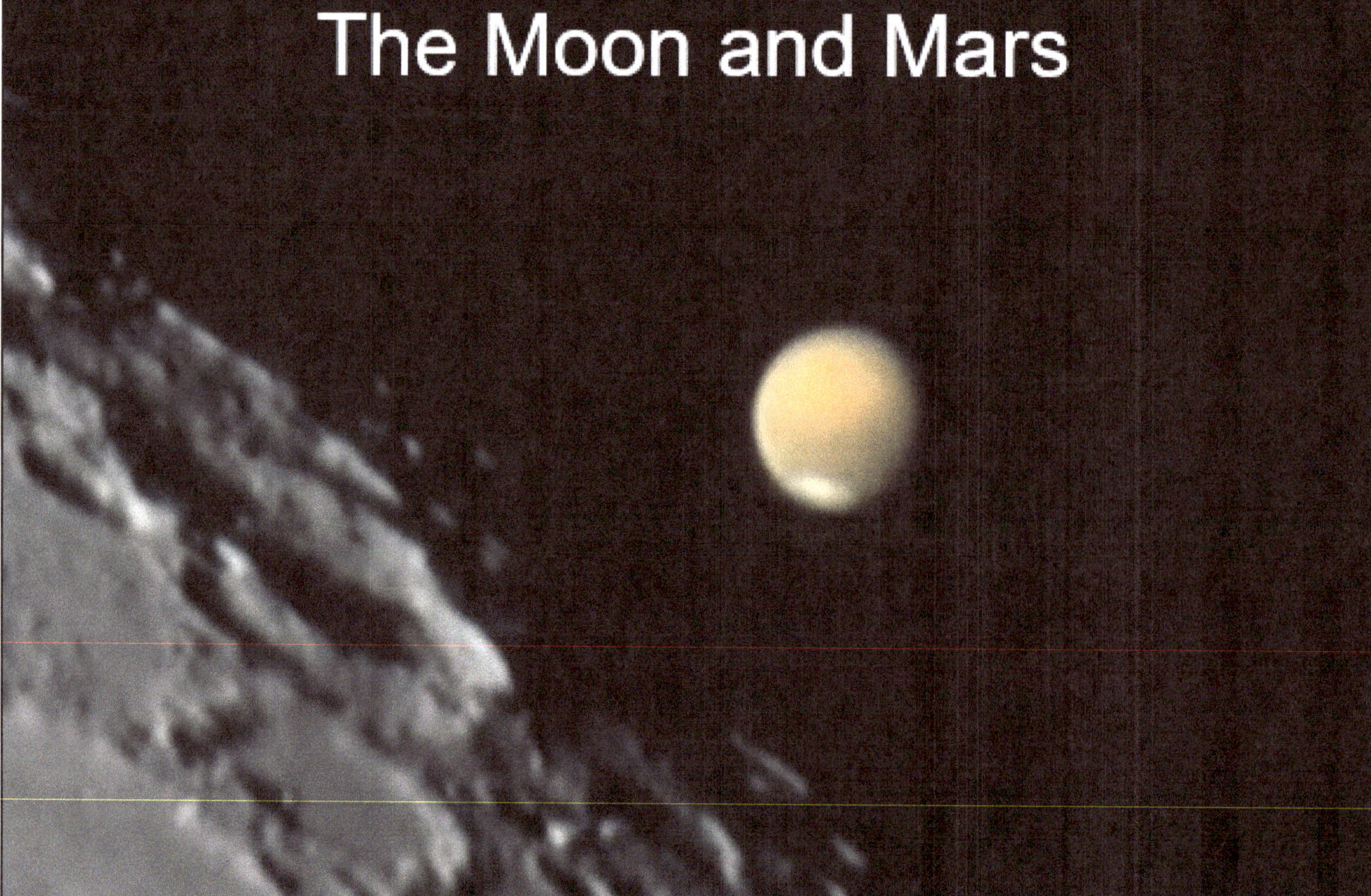

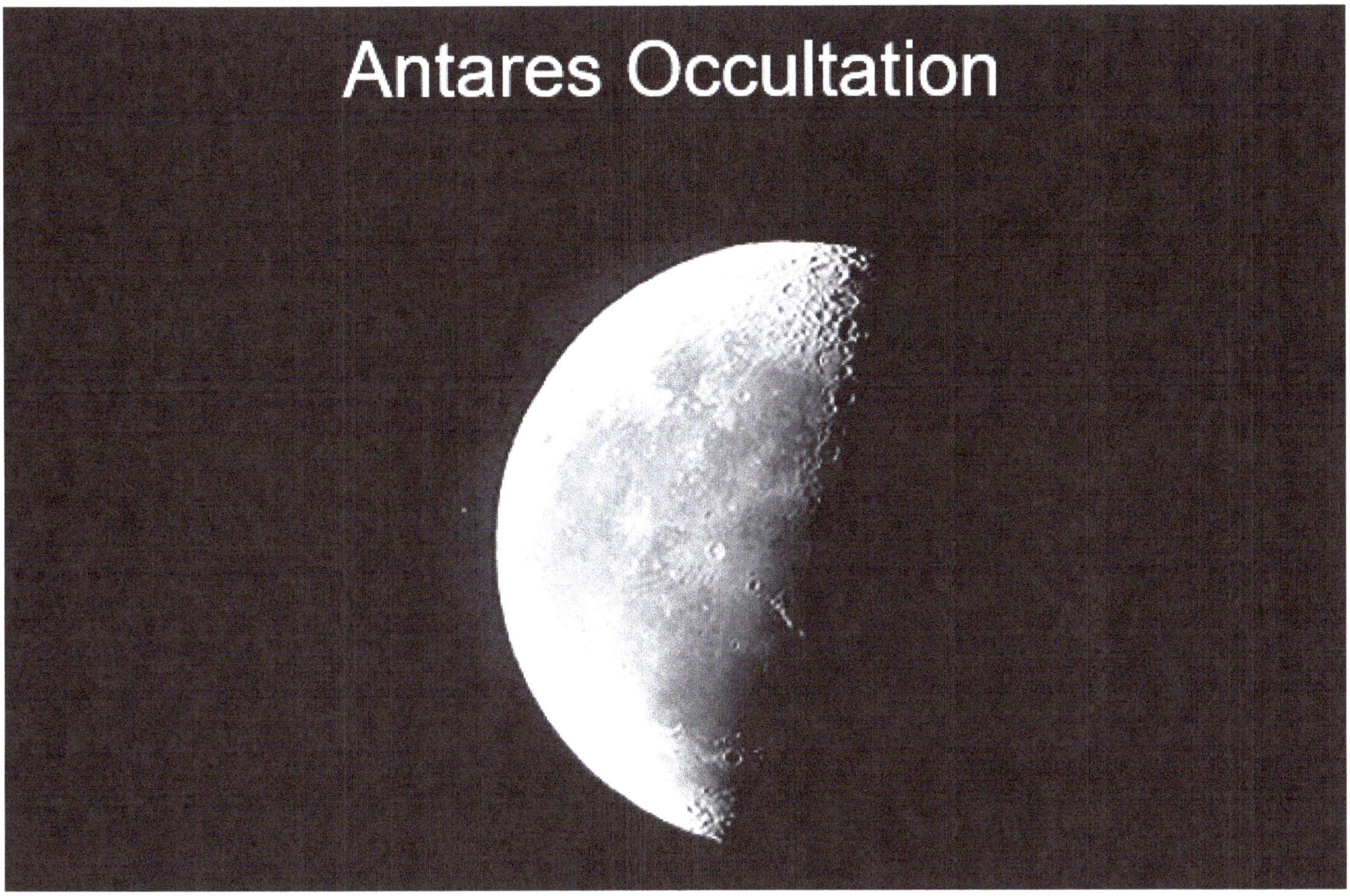

Antares Occultation

Antares is a red supergiant star, the brightest one in the constellation Scorpius.

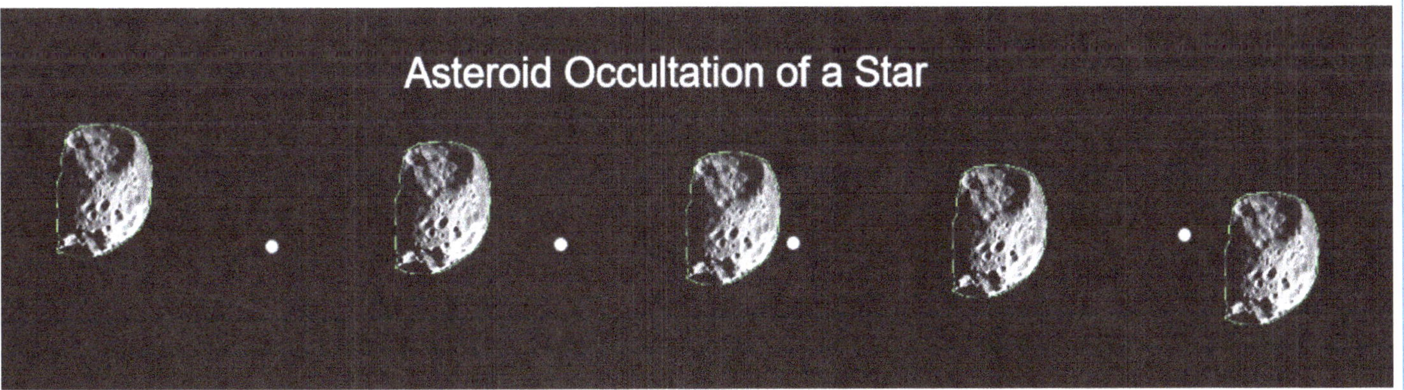

The Moon and Jupiter

The Moon and Saturn

A ***transit*** is the astronomical event that occurs when one celestial body appears to move across the face of another celestial body, as seen by an observer at some particular vantage point.

A transit is like a solar eclipse, but the foreground object is typically much smaller than the larger appearing body in the background. The only two planets that can transit in front of the Sun are Mercury and Venus because they are closer to the Sun than we are on Earth.

In the photo below, Mercury is the dot near the top of the Sun. The "smudge" to the lower right is a sunspot.

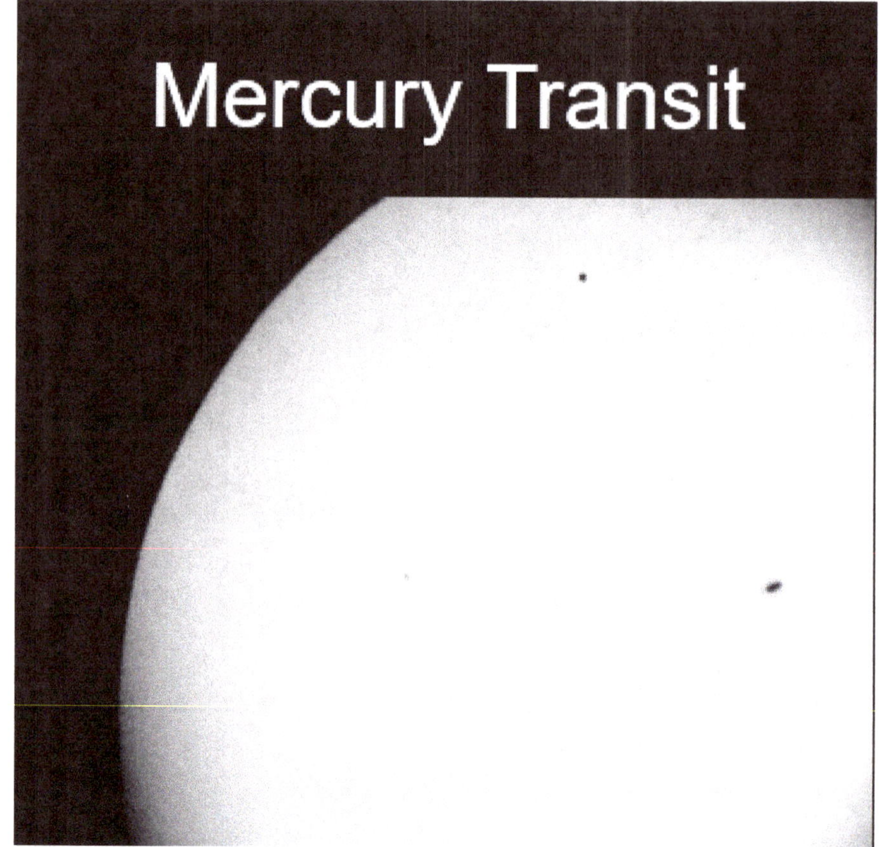

Transit of Mercury Across the Solar Corona

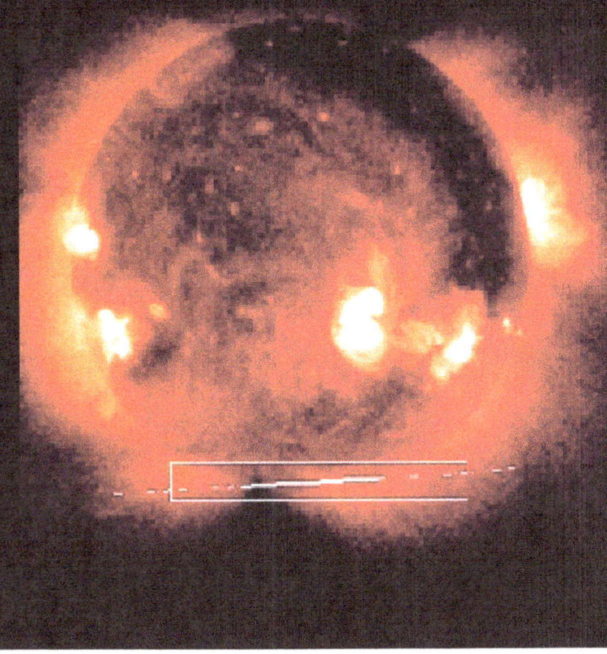

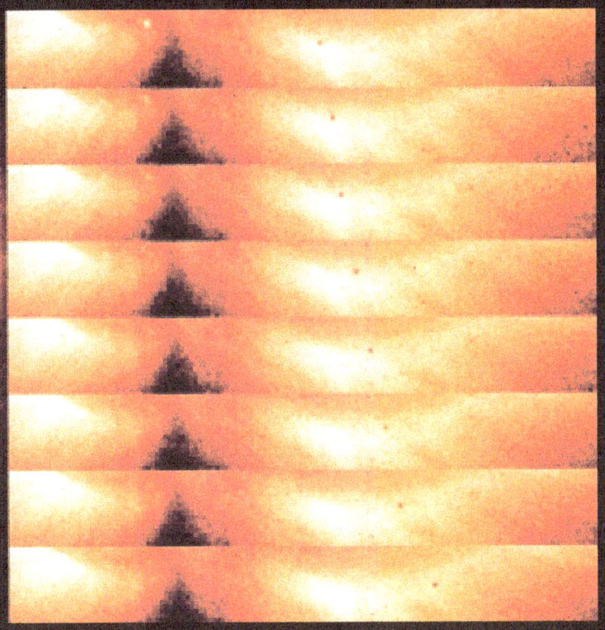

As a reminder, the solar corona is the Sun's atmosphere, and here we see a transit of Mercury in front of it just outside the limb edge of the Sun. To illustrate how rare it is, the last transit of Venus was on June 5-6, 2012, and was the last Venus transit of the 21st century. The prior transit took place on June 8, 2004. The previous pair of transits were in December 1874 and December 1882. The next transits of Venus will be in December 2117 and December 2125.

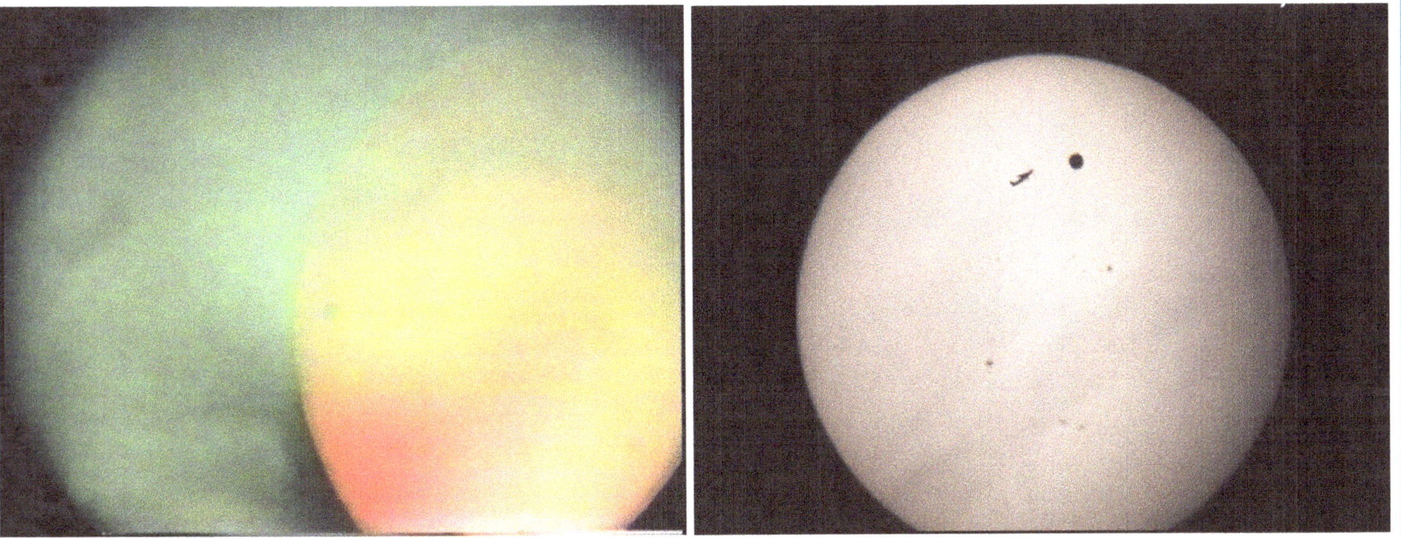

Using NASA's Solar Dynamic Observatory, this incredible image of the transit of Venus in 2012 was taken.

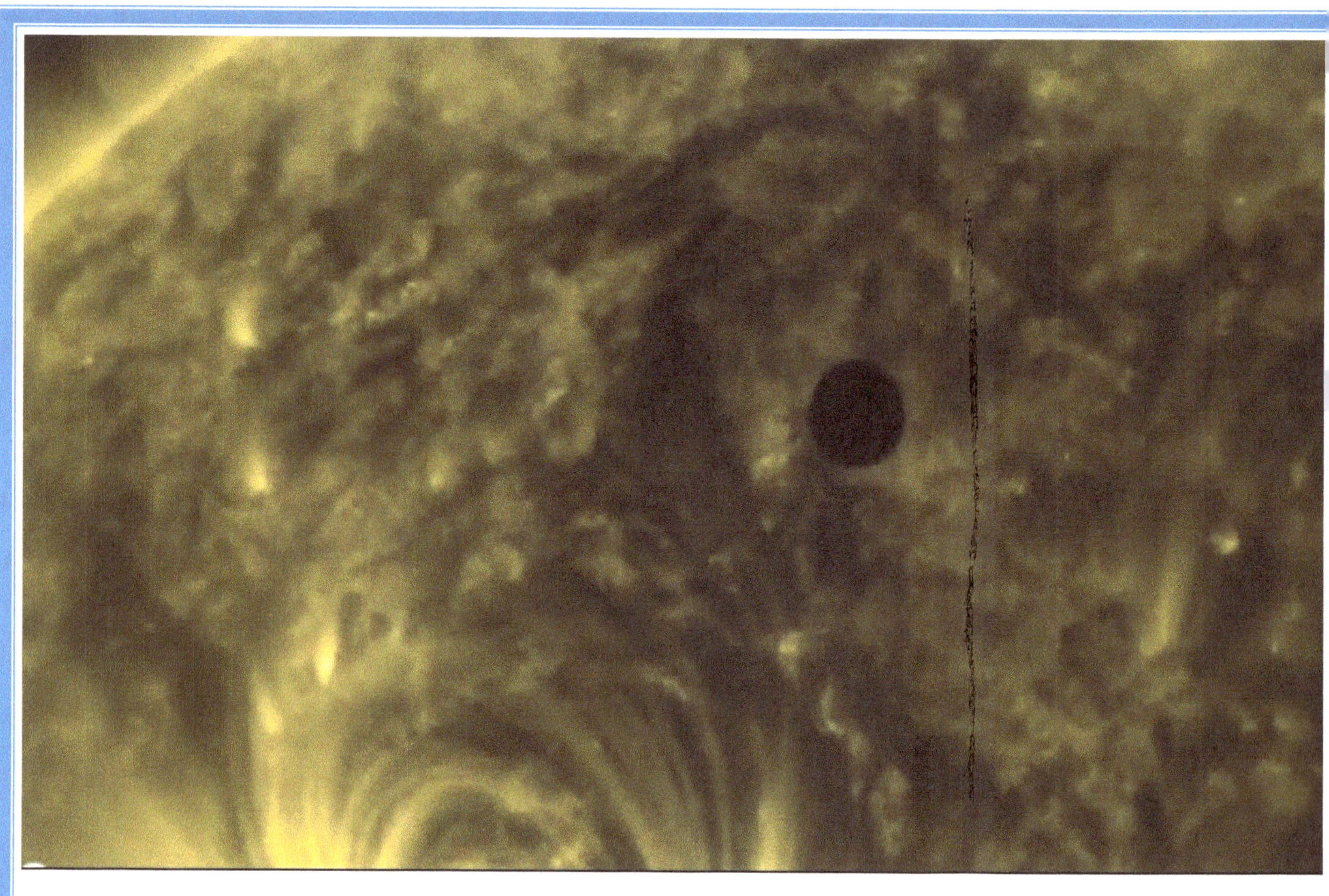

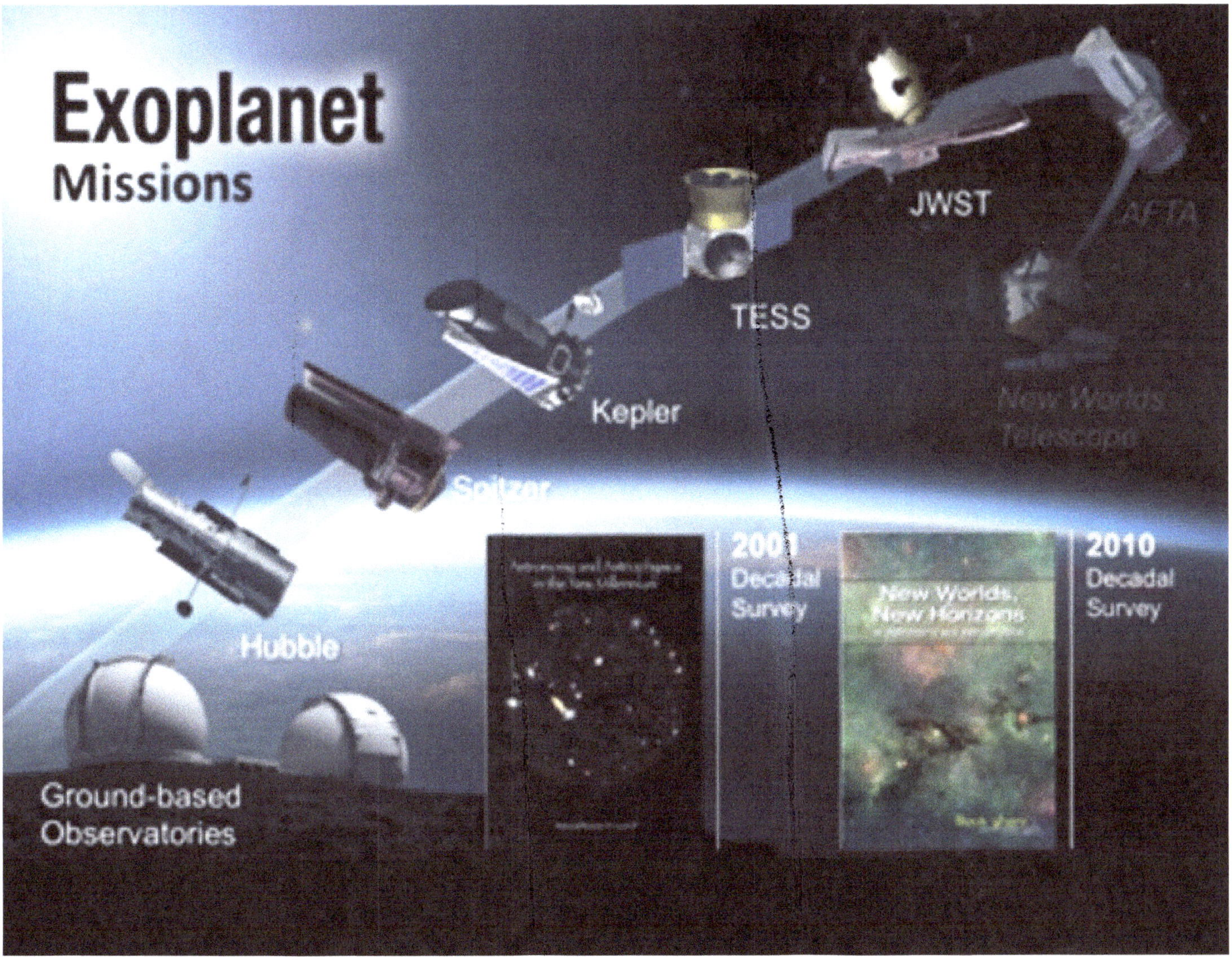

The primary way for the detection of extrasolar planets orbiting other stars beyond the Sun is called the transit method. Stars are so far away that it's almost impossible to "see" exoplanets orbiting them even with our largest telescopes, though recently we detected some giant exoplanets orbiting a couple of white dwarf stars with the James Webb Space Telescope using an occulting disc to block the bright starlight, as seen below.

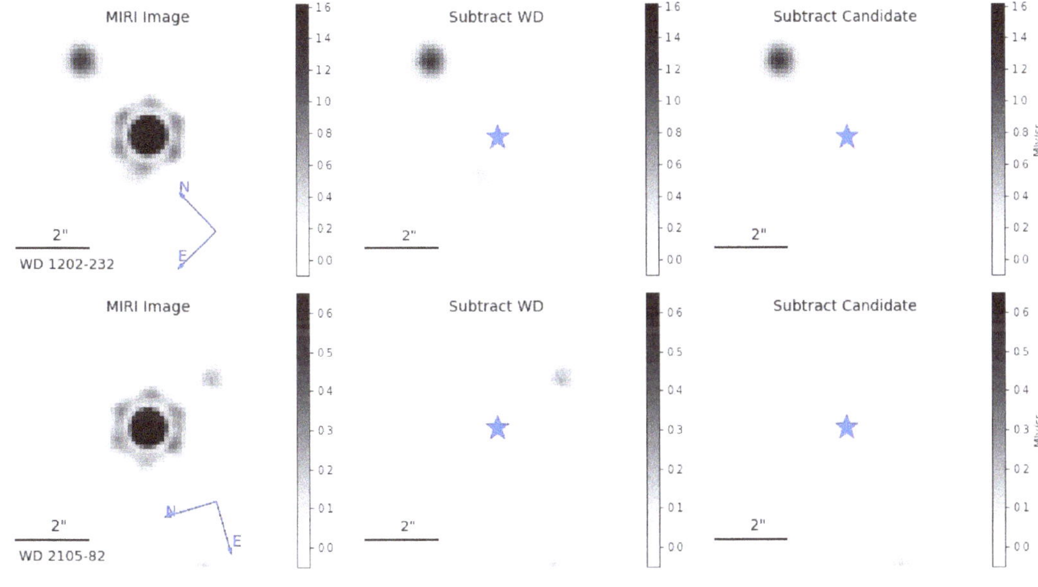

Credit: Sky & Telescope

For several years astronomers have been using the transit method to detect the presence of exoplanets. In spite of the drawback that it places limits on only those stars that have planets orbiting them in our line of sight, we've discovered thousands of exoplanet candidates. The way it works is that when a planet passes in front of its parent star, it physically blocks some of its light making the star appear slightly dimmer. Such small fluctuations in brightness can actually be measured using photoelectric photometry. As you can see in the illustration below, the change in brightness during the transit forms a classic bell curve.

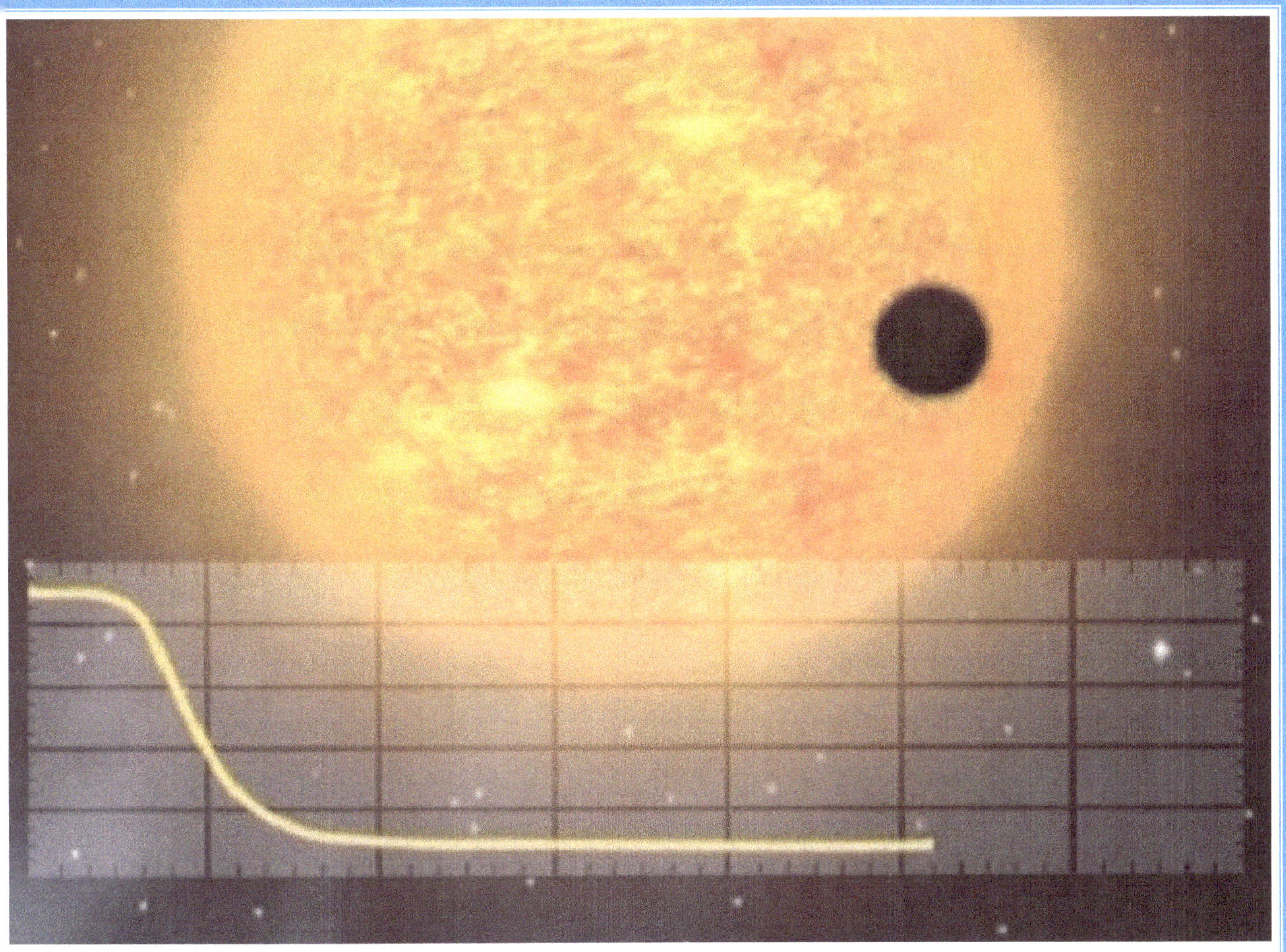

If you were the scientist in charge of this mission, knowing that the bell curve is all we can see, what would you need to see happen to provide some form of proof that the slight dimming is caused by a transiting exoplanet and not something like a star spot that comes and goes? You're right! Since planets, just like Earth, take a specific time to orbit their parent stars during their "year," we would expect to see the same bell curve occur again and again. A second time it happens would give us the period but a third time at the same duration would give us the "all clear" that it's most likely a good candidate for an exoplanet.

In general, transits are not uncommon. In the following picture, you are witnessing the transit of not one, but four moons of Saturn passing in front of their mother planet. The dark dots are shadows of the moons and mark areas where a solar eclipse is happening on Saturn. Of course, with temperatures plummeting below minus 225 degrees and an atmosphere rich in ammonia and methane, we don't think anything is alive seeing the eclipse. Titan, Saturn's largest moon and the 2nd largest moon in the solar system, is seen at the upper right. From the position of the sunlight, the shadow is off the edge limb of the planet.

Transit of 4 Moons of Saturn

Both natural and man-made objects can be seen to transit in front of larger appearing celestial ones, as seen in the incredible pictures below.

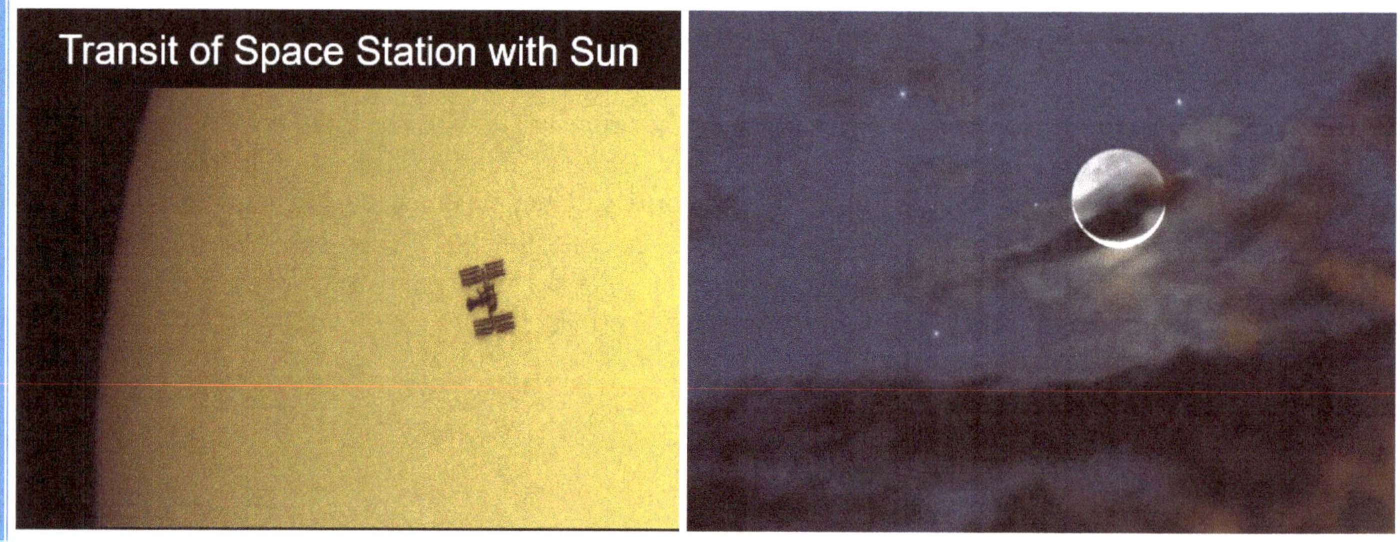

These earthly godfathers of
Heaven's lights, that give a
name to every fixed star,
have no more profit of their
shining nights than those
that walk and know not what
they are.
- William Shakespeare

When I Heard the Learn'd Astronomer

Walt Whitman

When I heard the learn'd astronomer,
When the proofs, the figures, were ranged in columns
  before me,
When I was shown the charts and diagrams, to add, divide
  and measure them,
When I sitting heard the astronomer where he lectured with
  much applause in the lecture room,
How soon unaccountable I became tired and sick,
Till rising and gliding out I wander'd off by myself,
In the mystical moist night air, and from time to time,
Look'd up in perfect silence at the stars.

The Lawrence Tree, 1929
Georgia O'Keeffe
Wadsworth Atheneum, Hartford

Daily Observation Log

Observer: _________________________________ **Date:** _________________

Time: _______________________ am/pm **Duration:** _______________________ min

Sky: 0 1 2 3 4 5 (circle one)   **Seeing:**   0 1 2 3 4 5 (circle one)

Constellation(s): _________________________________

____

____

____

Star(s): ____

____

____

Planet(s): ____

Object(s): ____

Phenomen: ____

Observational Method: unaided eye   paper tube   binoculars   telescope   (circle one)

Drawing:

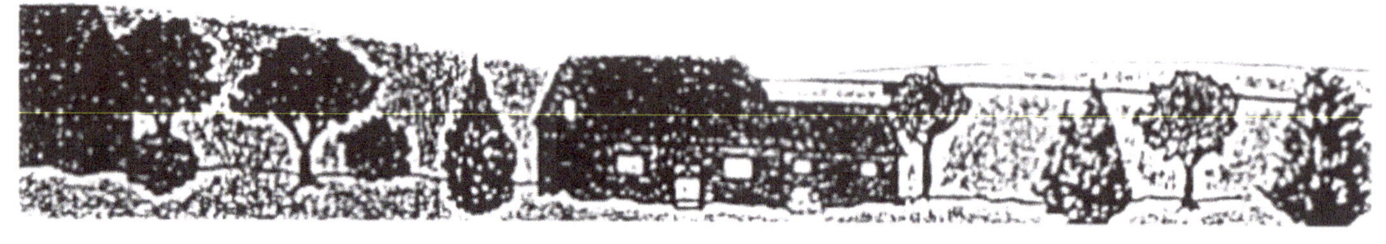

Instructions for Completing Daily Observation Log

Observer: Please print your full name

Date: Record current month/day/year (i.e. 01/08/2009)

Time: Record the time you began the observation and circle AM or PM

Duration: Record the total number of minutes you actually made your observation

Sky: Circle one number that best represents the sky from clear to completely overcast. 0 = clear; 1 = a few small clouds; 2 = partly cloudy; 3 = sky 50% cloud-covered; 4 = few breaks in clouds; 5 = completely overcast

Seeing: Circle one number that best represents the seeing conditions from excellent to poor. "Seeing" is a term used by astronomers to describe the steadiness of the atmosphere. One method of determining how steady or unsteady the atmosphere is, due to air currents and temperature changes, is by studying the brighter stars. Bright stars that appear to "twinkle" indicate turbulence in the layers of air in the atmosphere. Rate the seeing conditions on a scale of 0 for perfectly steady to 5 for stars that appear to "dance" in the sky.

Constellation(s): List any constellation you are able to identify in the night sky.

Star(s):   Write the name of each brightest star you are able to identify by consulting a star chart or atlas.

Planet(s):   Write the name of any planet you identify by referring to current data available giving its location.

Object(s):   Record the number and types of objects seen in the sky. Examples include meteors ("falling or shooting stars"), satellites, comets, asteroids, etc.

Phenomena: Any form of sky glow, such as aurora or the Milky Way, may be recorded

Observational Method: Circle the method of observation used. More than one per observation period can be utilized.

Drawing: Draw the moon phase (amount of sunlit portion) if visible. Also draw in anything recorded for that day's observation. You should draw in boundary lines separating different parts of the sky and include the direction abbreviated (i.e. SW) for each segment.

THE SUN

... INTERESTING TELESCOPE OBJECT AT 40x TO 70x BUT YOU MUST USE A SUN FILTER TO AVOID SERIOUS INJURY TO YOUR EYE. THE SUN SPOTS ARE EASY TO SEE

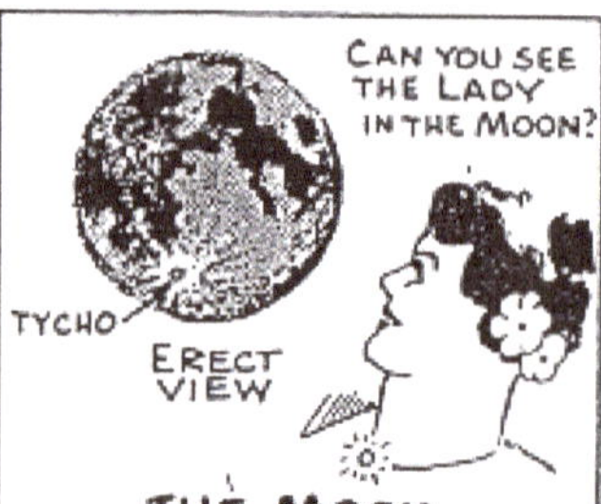

THE MOON

MAGNITUDE -12 WHEN FULL IS 190,000 TIMES BRIGHTER THAN FIRST MAGNITUDE STAR. CRATER TYCHO (TIE-CO) IS ON SOUTH SIDE - MOST PHOTOS ARE SHOWN INVERTED

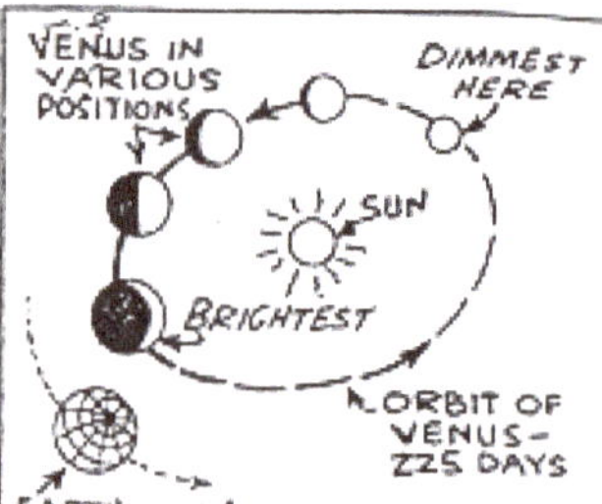

VENUS

LIKE ALL OF THE PLANETS, VENUS ORBITS AROUND THE SUN AND IS LIGHTED BY THE SUN. ON HER NEAR APPROACHES TO THE EARTH SHE IS BRILLIANT AT -4 MAGNITUDE

JUPITER

BIG JUPE IS THE EASIEST PLANET TO SEE -- ALWAYS BRIGHTER THAN -1½ MAG. HIS FOUR BRIGHTEST MOONS OF MAG. 6 SHUTTLE BACK AND FORTH, CHANGING NIGHTLY

SATURN

SATURN IS THE PRETTIEST PLANET. THE RINGS ARE SEEN PLAINLY AT 40x ALTHOUGH INVISIBLE WITH 7x BINOCULAR. WITH HIGHER POWER YOU MAY BE ABLE TO SEE CASSINI'S DIVISION

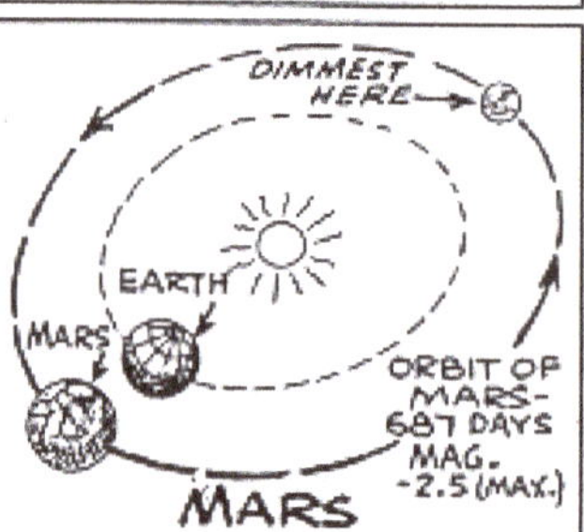

MARS

RED MARS MAKES A NEAR APPROACH TO THE EARTH EVERY OTHER YEAR, AND AT SUCH TIMES SOME SURFACE DETAIL CAN BE SEEN WITH TELESCOPES AT 200-300x

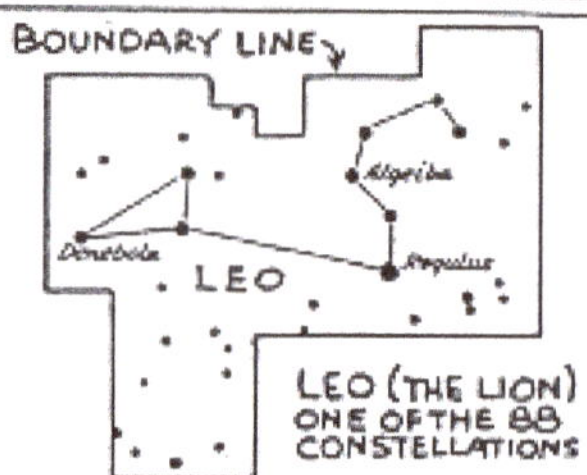

CONSTELLATIONS

A CONSTELLATION IS A GROUP OF STARS, USUALLY FORMING SOME KIND OF PATTERN OR "PICTURE." PROPERLY, A CONSTELLATION IS A SPECIFIC AREA OF THE SKY

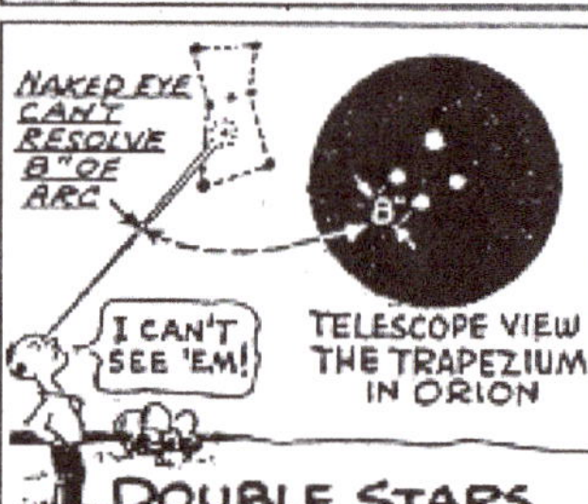

DOUBLE STARS

ONE OUT OF 15 STARS IS A DOUBLE OR MULTIPLE STAR AND ABOUT 500 OF THESE FROM 2 SECONDS TO 1 MINUTE OF ARC SEPARATION CAN BE "SPLIT" WITH SMALL TELESCOPES

VARIABLE STARS

A VARIABLE STAR VARIES IN BRIGHTNESS. THE CHANGE TAKES 2 DAYS (AVERAGE), MAKING THE V.S. A POOR "SHOW" OBJECT ALTHOUGH IDEAL FOR SYSTEMATIC STUDY

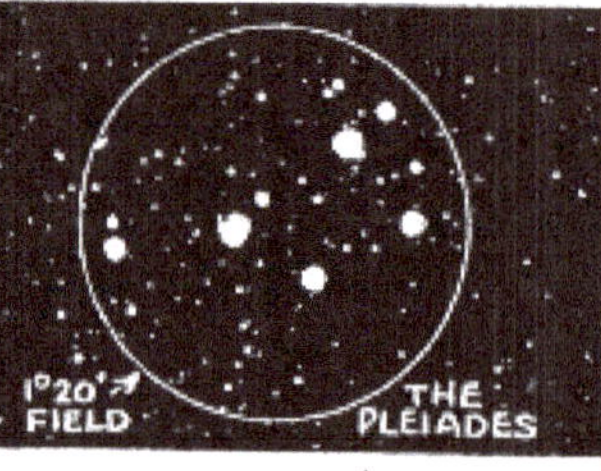

OPEN CLUSTERS

OPEN CLUSTERS OF STARS ARE A FAVORITE TARGET FOR THE TELESCOPE. 40 to 60x IS ENOUGH FOR MOST GROUPS. POPULAR PLEIADES CLUSTER IS A FINE BINOCULAR OBJECT

GLOBULAR CLUSTERS

A GLOBULAR CLUSTER IS A BALL OF STARS. INDIVIDUAL STARS ARE FAINT AND NEED 6" OR MORE APERTURE FOR RESOLUTION. M13 AND M22 ARE TWO BRIGHTEST

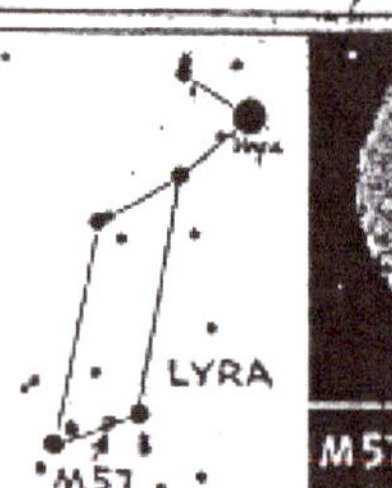

PLANETARY NEBULAE

PLANETARY NEBULAE ARE SO NAMED ONLY BECAUSE THEY ARE ROUND LIKE PLANETS. THEY ARE LUMINOUS GAS CLOUDS AND ARE A PART OF OUR GALAXY

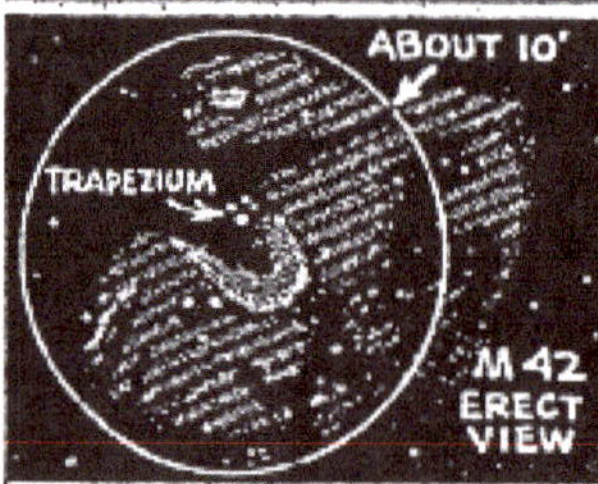

DIFFUSE NEBULAE

A LARGE DIFFUSE GAS CLOUD LIGHTED BY THE STARS IN ITS VICINITY IS KNOWN AS A BRIGHT DIFFUSE NEBULA. M42 IN ORION IS IMPRESSIVE, EASILY SEEN WITH ANY TELESCOPE

EXTERNAL GALAXIES

GALAXIES ARE COMPLETE STAR SYSTEMS LIKE OUR OWN GALAXY. ALL ARE VERY DISTANT. M81 SHOWN IS ABOUT AS BRIGHT AS A STAR OF 9th MAGNITUDE

NO.	TYPE	CONS.	M.
M44	OPEN CL.	CANCER	3.7
M41	OPEN CL.	CANIS MAJ.	4.6
M24	OPEN CL.	SAGR.	4.6
M31	GALAXY	ANDR.	4.8
M35	OPEN CL.	GEMINI	5.3
M13	GLOBULAR	HERCULES	5.7
M22	GLOBULAR	SAGR.	5.9
M8	DIFFUSE NEB.	SAGR.	—
M42	DIFFUSE NEB.	ORION	—
M57	PLANETARY	LYRA	9.3

MESSIER OBJECTS

FRENCH ASTRONOMER, CHARLES MESSIER, MADE UP THE FIRST LIST OF SKY OBJECTS OTHER THAN STARS (1784). ALL OF THE 103 M-OBJECTS CAN BE SEEN WITH SMALL TELESCOPES

Constellations

Directions: Using your pen or pencil, connect the dots (stars) to form the stick figure patterns of the stars making up each constellation.

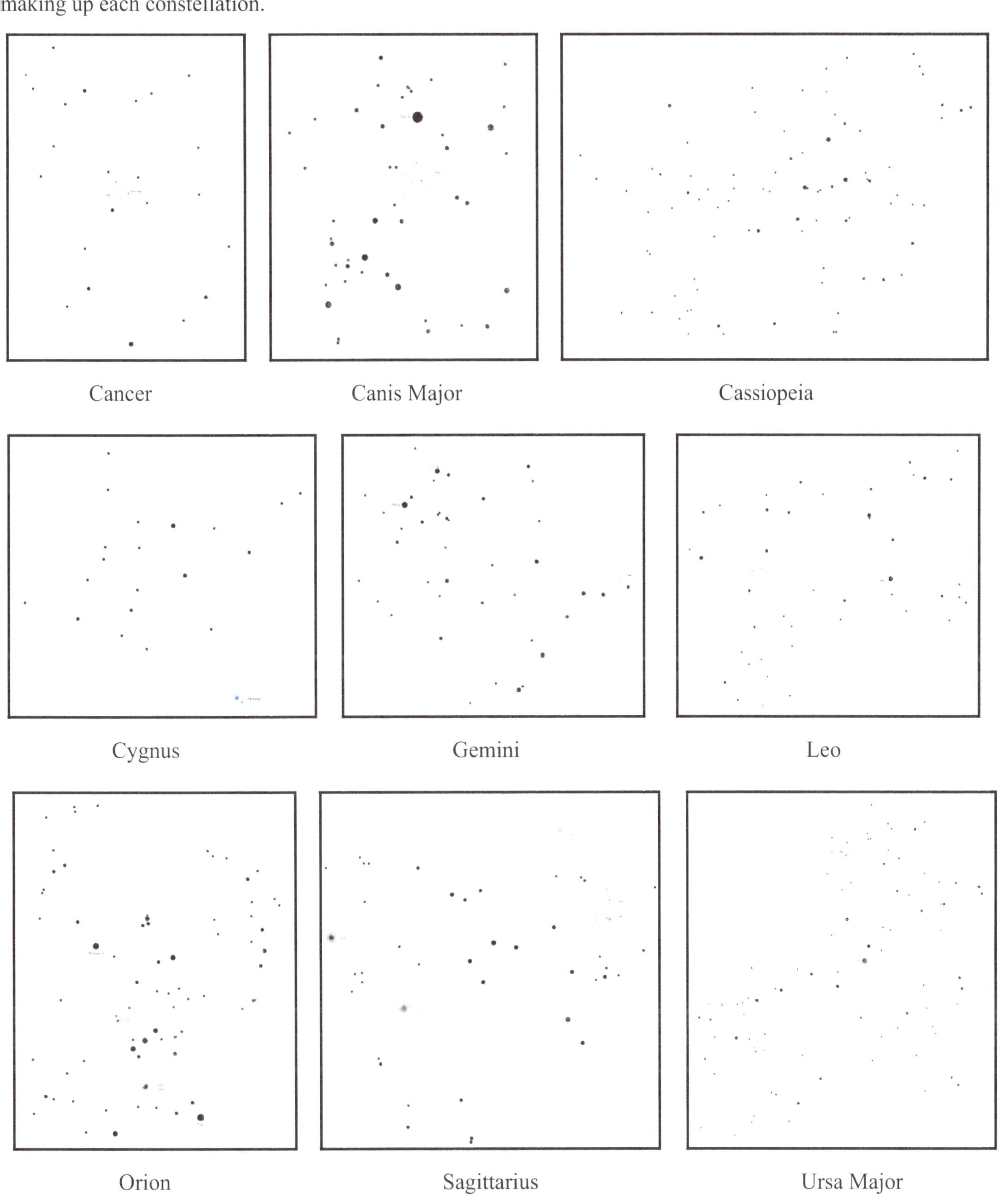

VIEWING SUN SPOTS

Sunspots are areas of particularly strong magnetic forces on the Sun's surface that appear darker than their surroundings because they are cooler.

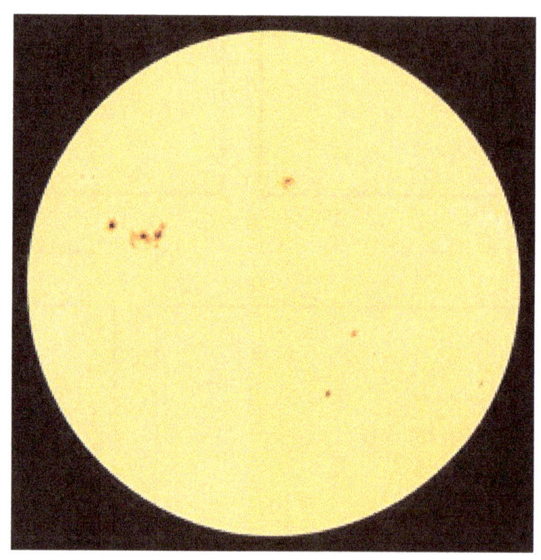

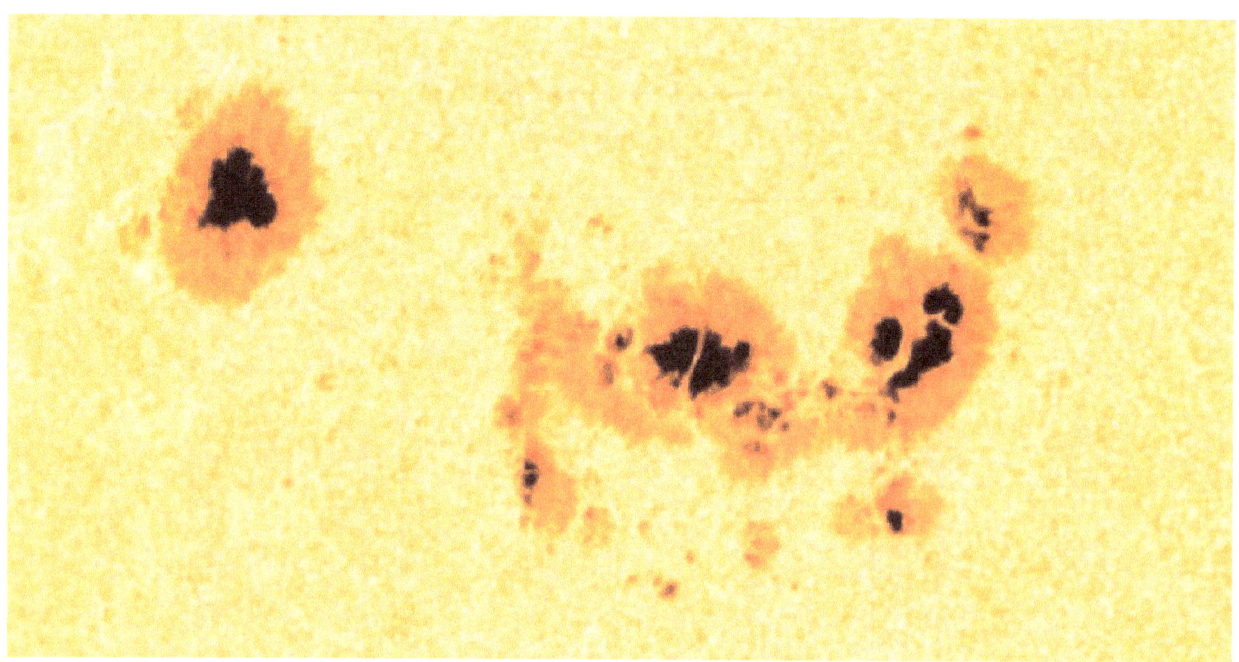

Record Your Daily Observations of Sunspots

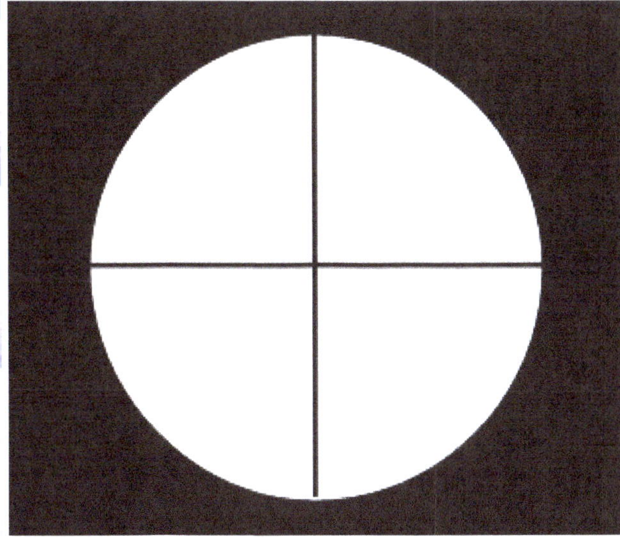

Date ______________ Time ______________

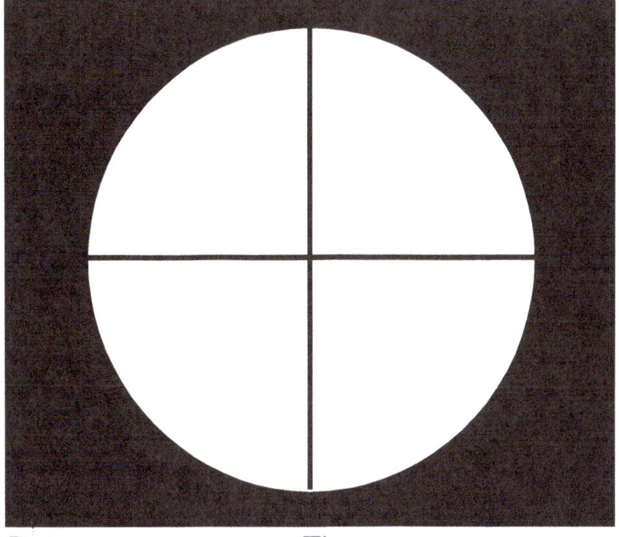

Date ______________ Time ______________

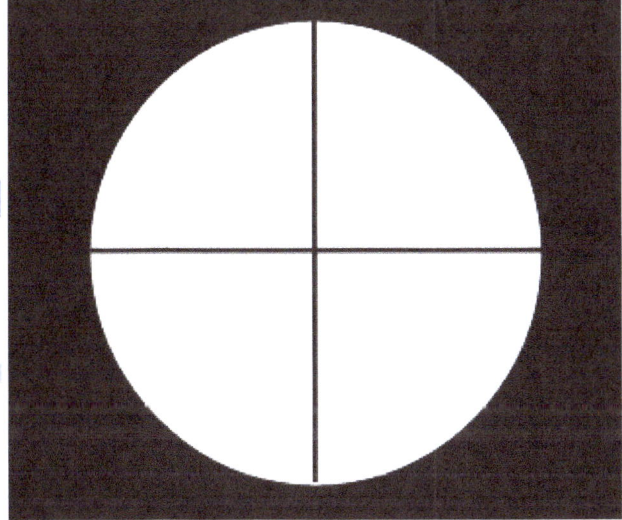

Date ______________ Time ______________

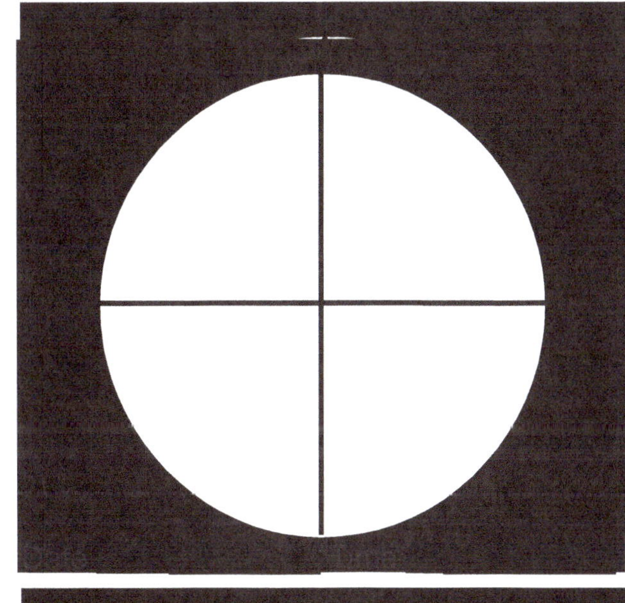

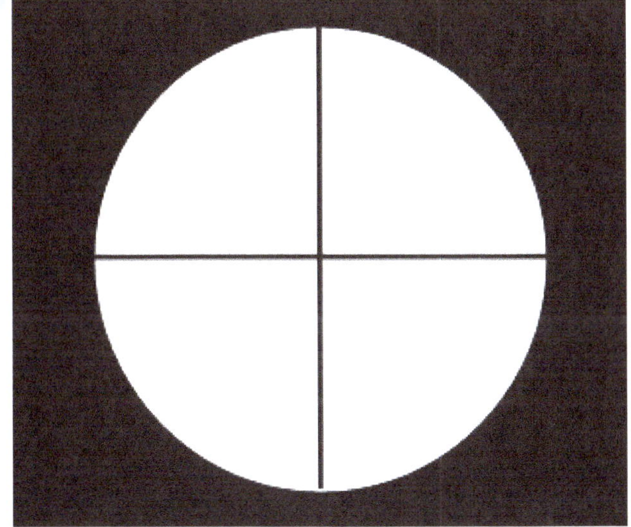

Date ______________ Time ______________

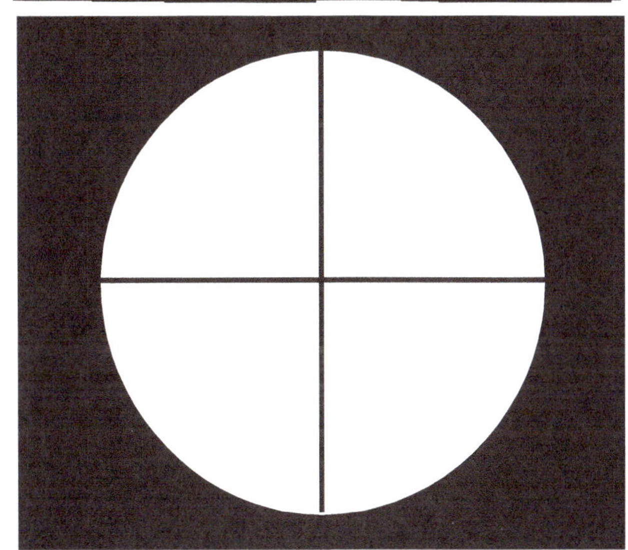

Date ______________ Time ______________

FACES ON THE MOON

Directions: Draw as many figures of people and animals you can find on the view of the Moon below and label them at the numbered list at the bottom of the page.

1. ____
2. ____
3. ____
4. ____
5. ____
6. ____
7. ____

Motions of Earth Through the Universe

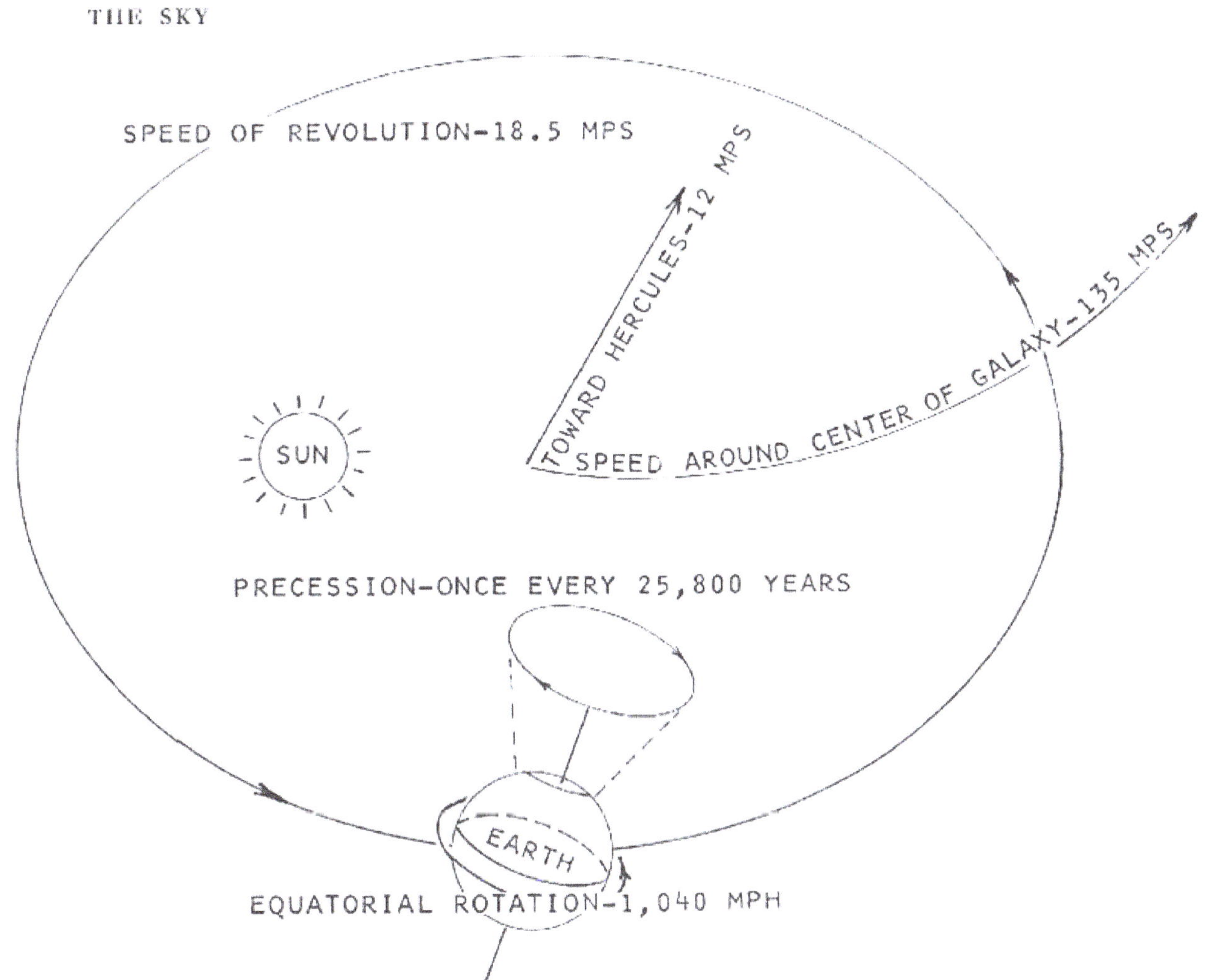

Some of the earth's motions (not drawn to scale).

Can you guess what the 6th motion might be?

Thank you and hope you learned some new things about these awesome celestial events.

About the Author, Kevin Manning

Brief Bio

Consultant with NASA
Chandra X-Ray Observatory
- **Harvard-Smithsonian Center for Astrophysics**

Wright Fellow at Tufts University
Einstein Fellow on Capitol Hill in Washington, DC
- **NASA Headquarters**
- **US House of Representatives**
- **US Dept of Energy Office of Science**

Brookhaven National Laboratory

Noteworthy Workshops

- **Tufts University**
- **State University of New York at Stony Brook**
- **National Science Teachers Association's National Convention**
- **American Association for the Advancement of Science Breakfast with Scientists**
- **National Parks Service**

You can reach Kevin via email at kevin@lookuptothestars.com